PERIL IN THE PONDS

PERIL IN THE PONDS

Deformed Frogs, Politics, and a Biologist's Quest

JUDY HELGEN

University of Massachusetts Press
Amherst and Boston

Printed in the United States of America

LC 2012007994
ISBN 978-1-55849-946-1 (alk. pbk)
ISBN 978-1-55849-945-4 (alk. cloth)

Designed by Sally Nichols
Set in Adobe Garamond Pro and Myriad Pro
Printed and bound by Thomson-Shore, Inc.

Library of Congress Cataloging-in-Publication Data

Helgen, Judith Cairncross.
Peril in the ponds : deformed frogs, politics, and a biologist's quest / Judy Helgen.
p. cm.
Includes bibliographical references and index.
ISBN 978-1-55849-946-1 (paper : alk. paper) — ISBN 978-1-55849-945-4 (library cloth : alk. paper)
1. Frogs—Abnormalities—Minnesota. 2. Frogs—Habitat—Minnesota. 3. Wetland ecology—Minnesota. 4. Water—Pollution—Minnesota. 5. Indicators (Biology— Minnesota. 6. Helgen, Judith Cairncross. 7. Biologists—Minnesota—Biography. 8. Environmental protection--Minnesota. 9. Environmental policy—Minnesota. 10. Minnesota—Environmental conditions. I. Title.
QL668.E2H42 2012
597.8'9—dc23
2012007994

British Library Cataloguing in Publication data are available.

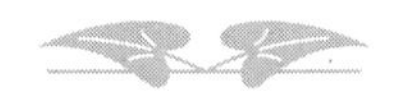

For all who love nature and try to protect it

For children who observe the living world and ask us to care

For Lila and Oskar, Lucas and Nathaniel

For Verlyn

CONTENTS

Acknowledgments ix

Introduction 1

1. The Call 7

2. The Frog Champion 28

3. Learning Curves 42

4. Wading in and Listening 68

5. The Peril Widens 77

6. Science in the Public Eye 94

7. Government to the Rescue? 109

8. The Quest 137

9. Imperiled Frogs 149

10. Bureaucratic Strangulation 172

11. The Wrecking Ball 188

Epilogue: How Are the Frogs? An Update 200

References 225

Index 241

ACKNOWLEDGMENTS

So many people have helped along the way I can't possibly name them all. First, I thank my comrades in wetlands and frogs, especially Mark Gernes, but also many people who worked with us at the Minnesota Pollution Control Agency: wetland field assistants Joel Chirhart, Cade Steffenson, and Kyle Thompson; intrepid frog workers Dorothy Bowers, Jeff Canfield, Drew Catron, Sue Kersten (Vanden Langenberg), Pam Schense, Phoebe Vanselow, and others. I thank many MPCA staff who tolerated and supported our work: Duane Anderson, Greg Gross, Bob Murzyn, Ralph Pribble, and MPCA's patient librarians, Kathy Malec and Helena Peskova; the reference librarians at the University of Minnesota; my Minnesota frog teachers Carol Hall, Dave Hoppe, Bob McKinnell, and John Moriarty, as well as Gary Casper in Wisconsin; the students in the Frog Group who told us all about the frogs: Jack Bove, Reta Bove (Lind), Jeff Fish, Ryan Fisher, Betsy Kroon, Becky Madison Pollack and husband Nick Pollack, and Guthrie Swenson; the teachers, Tom Fish in St. Peter, Cindy Reinitz and administrator Dee Thomas from the Minnesota New Country School in Henderson, Gail Thovson in Litchfield. Also Mary Kay Lynch for ardent help teaching our wetland volunteers. I thank the late Don Ney for donating his land for the Ney Nature Center and for sharing his stories with me.

A special thanks to Art Straub, who keeps the faith by teaching kids about nature and tracking the wildlife in the Minnesota River valley, and to the inspiration, support, and commitment of the late member of the Minnesota House of Representatives Willard Munger. Many scientists pitched in: Jim Burkhart, Perry Jones, Kathy Lee, Tim Kubiak, Mike Lannoo, Ed Little, Carol Meteyer, Mike Thurman, and a host of others who openly shared ideas and offered help. Special thanks to reporters Dennis Lein, Mary Losure, Sarah Malchow, Tom Meersman, Dean Rebuffoni, Ken Speake, and others who told the story fairly. And to the hundreds of concerned citizens, thank you for speaking up about your frogs—people like the Bocks, Joy Jacobsen, Audre Kramer, the Tousleys, and many others who had the courage to let government scientists survey their frogs and take samples on their properties. I thank Anne Dubuisson Anderson for early editorial assistance, Scott Edelstein for savvy advice on publishing, and Susan Corey Everson for her helpful comments.

I couldn't have written this book without the attentive and critical members of my writers' groups: Tom Anderson, Val Cunningham, Patti Isaacs, Sue Leaf, Judy Krauss, Gayla Marty, Sue Narayan, Darby Nelson, Doug Owens-Pike, Kate Quinlan, Pam Schmid, and the teachers at the Loft Literary Center who have helped me begin to learn to write. I'm grateful for the thorough critiques by two peer reviewers and by my editor at University of Massachusetts Press, Brian Halley, and for the helpful review of the introduction by author Ross Gelbspan. Their comments have helped make this a better book. Finally, I thank my ever tolerant and beloved husband, Verlyn Smith, who never knew that frogs would continue to occupy my days even in retirement, and my sons Erik and Steve and their families for moral support.

PERIL IN THE PONDS

INTRODUCTION

I knelt on the ground in my oversize rubber waders and peered into the metal pan. Its water danced with small creatures we'd just netted from the pristine-looking river below. A variety of immature insects swam about: armored dragonflies and pebble-cased caddis flies, dark-bodied beetles and bugs. Other invertebrates—tiny crustaceans and elegantly spired snails—crawled along the bottom of the tray. I looked down at the San Marcos, its clear blue water sparkling in the midday sun. Who knew it harbored such a diversity of life?

A thirty-something mother, I was taking this course in aquatic ecology as part of my quest for a new direction in life, one I hoped could support our young family. A dozen years had elapsed since my first graduate work in zoology. I knew I'd be starting all over. Would I return to my early interest in cell and molecular biology—or try environmental law? My husband held a temporary teaching position at a small college in Texas for the academic year. The coming summer we would have to move, but where?

That day our aquatics class had also dipnetted the murky water of a river known to be heavily polluted. The results were disappointing: a few squirmy worms, some red insect larvae, a snail or two. Something shifted in me then. Why this stark difference? Seeing the lively organisms we'd taken from clean water and the sluggish poverty from the polluted river pushed me onto a new path. This was it. I wanted to protect aquatic organisms from water pollution. To accomplish this, I needed advanced training and decided to enter a graduate program at the University of Minnesota, where we'd be closer to our extended family.

By the end of the academic year we packed up and headed north from Texas to Minnesota, where I started a PhD program in zoology. With abundant lakes and streams, Minnesota would be a fine place to study aquatic ecology and, hopefully, secure a relevant job.

Viewing that tray of wiggling water creatures in the mid-1970s, I had no idea that twenty years ahead I would be analyzing tiny aquatic animals to measure the biological health of wetlands; that the EPA would support my work as a research scientist in a state pollution control agency; that I'd be wading into muddy-bottomed swamps to dipnet the invertebrates and let them tell me if the water was polluted or not. I had no idea then that over the next two decades biologists would document the extinctions of several species of frogs from areas in Australia and Costa Rica. Nor did I know that many other global populations of frogs and amphibians would decline sharply.

I couldn't have foreseen that, in the future, I'd be crouching down on damp grass to examine young frogs I'd netted from wetlands; that I'd be repeatedly shocked by what I saw: a completely absent leg, a stump instead of a limb, two feet branching from one joint, a missing eye, or an extra leg flopping uselessly to one side. And I couldn't foresee that much of my work in the 1990s, aimed at protecting wetlands and understanding why frogs had deformities, would be controversial.

As early as the 1960s, warning signals of human-caused damage to the environment and to its living creatures were widespread and attracting attention. It wasn't just the disastrous oil spill that polluted Santa Barbara's beaches in California or the notorious Cuyahoga River in Ohio, where a half-foot floating slick of waste oil repeatedly caught fire and burned for days at a time. It was robins quivering and dropping dead after eating DDT-laden earthworms. It was reports that radioactive fallout from nuclear bomb tests in Nevada appeared in the teeth and bones of children who drank milk from cows that had grazed on contaminated grass. Shocking pictures of malformed babies born with deformed or missing limbs caused by the drug thalidomide given to pregnant women alerted the public that pharmaceuticals could be hazardous to the fetus.

In 1962, the publication of Rachel Carson's book *Silent Spring* focused and intensified the general public's growing concerns about the health of the environment. Carson exposed the dangers of commonly used pesticides and the appalling lack of government testing and regulation. She taught us how chemicals pass insidiously through food webs from plants to plant feeders, like insects and worms, to consumers such as robins, fish, eagles, and humans.

Carson's warnings remain relevant even today: only a fraction of the thousands of chemicals in current use have been tested for biological effects. The identities of many other chemicals, for instance, those added to pesticide mixtures, are kept secret because they are considered "confidential business information." In addition, tests for chemical toxicity are not typically conducted for long enough periods of time to check for developmental effects in the test organisms. More than two hundred chemicals have been monitored in the tissues of humans (CDC 2009).

By the 1970s, later called the Decade of the Environment, the first Earth Day drew twenty million people onto the streets, and ordinary citizens openly urged the government to rein in pollution. The US Environmental Protection Agency (EPA) had been established and the Clean Water Act enacted, along with other major laws aimed at protecting the environment. With federal agencies strengthened by the new legislation, positive changes ensued across the country. Rivers no longer burned, hatches of mayflies, absent for years, returned, fluttering over cleaner rivers and lakes. One long-term employee of my state's pollution control agency, who helped communities develop wastewater treatment facilities, recently told me that back then anything he did made a difference.

During the 1970s and 1980s, pollution control meant regulating end-of-pipe discharges of sewage and certain toxic chemicals into rivers. Water-born pollutants and some physical factors, such as temperature and acidity, were routinely monitored in streams, but not their aquatic organisms. At the onset, many water quality staff in pollution control agencies had backgrounds in civil engineering and biologists were rare. Yet a central goal of the Clean Water Act is to restore and maintain not only the physical and chemical integrity but also the *biological integrity* of the nation's waters (Wellock 2007).

Over time, the EPA promoted a new paradigm for monitoring water pollution, one that necessitated analyzing the condition of the biological communities inhabiting surface waters. What do the types and assemblages of aquatic organisms—fish, aquatic insects, and plants—tell us about the water's health? Do existing chemical standards for preserving the water quality of streams and lakes actually protect their aquatic species? EPA suggested that biological monitoring could be used more effectively than testing single chemicals to determine which state waters are the most polluted. In the 1990s some states, Ohio for instance, were documenting pollution in rivers by analyzing the health of fish communities.

Then the EPA began to encourage states to test the water quality of wetlands, not just streams, by examining their water-dwelling animals and water-dependent plants. *Wetlands?* The idea did not sit well with some state pollution managers, nor does it today. Monitoring chemical pollution in wetlands, let alone examining wetlands organisms, was not a part of most water quality program plans. In many states wetlands weren't even officially considered to be "waters of the state."

By the 1990s, as a fledgling government scientist who waded into wetlands and handled hundreds of deformed frogs, I faced many obstacles. My agency seemed unprepared to tackle such unexpected and politically inflammatory problems as pollution in wetlands, let alone deformed frogs first found by school children. But the same state government agency, formed in 1967, had worked assiduously to stop the flow of raw sewage into rivers, reduce the release of some harmful chemicals, and keep Lake Superior free of toxic mining waste. What happened during the 1990s that seemingly slowed government's environmental activism?

At the time, I was barely aware of efforts that were underway in Congress to undermine science; that in 1995 the Republican House (under Speaker Newt Gingrich) had dismantled the federal government's highly respected science advisory panel of more than one hundred scientists who wrote nonpartisan reviews of science issues for Congress. Hearings on "scientific integrity" were held, giving fringe scientists a platform, as if science itself was put on trial (Mooney 2005). Environmental regulations were maligned as a "costly and unnecessary government interference," in industry's view.

I didn't grasp the full implications of this movement. If I had, I might have better understood some of the challenges I faced with other government scientists when we promoted biological monitoring for wetlands and investigated the deformities in frogs. I would have better understood why our inclusion of pesticides among several possible causes of abnormalities in frogs would be called a "witch hunt" or having a case of "Rachel Carson syndrome."

The assault on science continued into the twenty-first century (Mooney and Kirshenbaum 2009). In the G. W. Bush administration (2001–2009), almost unattainable burdens of proof were placed on scientific findings that supported government regulations before action could be taken (Shulman 2006; Mooney and Kirshenbaum 2009). Some scientists' findings were repressed (Union of Concerned Scientists 2008). The established science of climate change was particularly targeted (Gelbspan 2004, 1997) and remains under fire today. And today, major environmental laws, such as the Clean Air and Clean Water Acts, are in jeopardy.

The quest for answers to the causes of deformities in frogs, however, was more hampered by what we didn't know about the biology of developmental abnormalities than by the politics of the 1990s. Scientists acknowledged gaps in basic knowledge even though great strides were taking place in developmental biology at the fundamental, molecular level. What disrupts the development of legs in amphibians (or in humans for that matter), causes missing eyes, or deforms internal bony structures, such as the spinal column and pelvis? How can a limb grow out of the side of a frog's abdomen, or an eye be found in its throat? Could this be caused naturally or by chemicals or ultraviolet light? We faced many unknowns while several hypotheses were proposed. I couldn't answer a frequently posed question: Might people be harmed too?

Researching chemical causes of deformities generated controversy and raised more questions. What chemicals are in the environments of frogs? How could we find out, when some of those chemicals were protected as trade secrets? Which chemicals had been tested to see whether or not they cause developmental abnormalities? Natural causes—such as parasites and predators—had an appeal and gained some scientific support. Over time, the mystery of the deformed frogs deepened, as this book will show.

Peril in the Ponds is a story about engaged students, alarmed citizens, an environmentalist legislator, and passionate scientists who tried to explain a startling problem: deformities in frogs. This story shows some of the political and scientific challenges faced by environmental biologists both within and outside of government and the concerns expressed by people who wanted action and answers. I guide readers into the natural world of wetlands—one of my favorite places—and inside the fascinating lives of frogs, which, surprisingly, share with us some common, biological needs for health and survival.

My narrative concludes with an epilogue that updates the current status of wetlands along with a review of some of the more recent research on malformed frogs. Abnormal frogs continue to emerge in significant numbers from small ponds all over the United States, Minnesota included. Recent decisions by the US Supreme Court have stripped away government protections from the very type of isolated wetlands that many kinds of frogs need to reproduce safely. Such wetlands include a diversity of ponds from seasonally wet pools to the more permanent depressions of the Prairie Pothole region in the Midwest. Wetlands losses continue even today.

Frogs and wetlands remain in peril. This is their story, too.

A Note to Readers

Throughout this book I use the terms *deformity, abnormality,* and *malformation* as meaning the same thing: development gone awry. However, in scientific circles, the proper term is *malformation.* In addition, I use the word *pond* to indicate the broad range of wetlands that frogs require, such as vernal pools, forested wetlands, permanent and semipermanent emergent, vegetated wetlands and marshes, and natural ponds. Also, nontechnical descriptions of the deformities are used.

CHAPTER 1

THE CALL

Her voice quavered over the phone as she described a hellish scene: frogs with stumps of legs; frogs missing a leg; frogs with twisted joints; some with extra legs that couldn't move. A frog with one eye. "Half of these frogs have something wrong. They look really pathetic." She paused. "We need help."

It was mid-August of 1995 when I first talked with Cindy Reinitz, a teacher who'd taken her students on a nature walk around a pond near Henderson, Minnesota, a community of nine hundred located close to the Minnesota River south of St. Paul. The town was founded in 1852 as a trading center, and today some of its old brick buildings are listed in the National Register of Historic Places.

Listening to Cindy, I held my breath. Another nightmare in a wetland, I thought, remembering the deformed frogs reported to me two years earlier by a woman who lived in Granite Falls in west central Minnesota. Half of the forty frogs she and her grandchildren had caught in her backyard that summer had deformities, she'd said, expressing concern for the children's safety. When I visited there, her husband, a burly tree cutter, revealed his fears that whatever harmed the frogs might drive them from their home. "This is the best place in the world to live," he told me, waving his big hands almost defensively at the woods and toward the dark waters of the Minnesota River flowing close by.

I knew that ponds were already in peril from pollution and regulatory neglect. And over the decades, thousands of them had been drained to create farmland in rural areas. But now, with Cindy's alarming report, I wondered, had ponds become too perilous, even for frogs? My mind began to spin. What agents could possibly infiltrate wetlands and disturb amphibian development? Photos of limbless frogs would soon raise warning flags. Were humans also in danger?

I doubted that my agency's administrators would agree to yet another frog project. When the deformed frogs found near Granite Falls were reported to me in 1993, I managed to secure emergency funding from the EPA to investigate the situation the following year. Our bosses went along with it. I was starting to research biological indicators for assessing wetlands health and not getting much support at the agency. Might deformed frogs emerging from ponds help shift their attention to wetlands?

The next year (1994) we returned to Granite Falls, collecting and examining hundreds of frogs around the area, including those emerging from the wetland where we'd seen deformed frogs the previous fall. Not one single frog had a deformity. We uncovered multiple sources of possible pollution—transformer oils, a rotten petroleum pipeline, and toxic metals (Gernes and Helgen 1997). And later on, state fisheries biologists measured some toxic chemicals in fish taken from the Minnesota River nearby. But all the frogs looked normal. I suspected that my managers were as puzzled by this as I was. Or perhaps they saw me as overreacting to an aberrant report of deformed frogs. An isolated case.

Now, with this new report from the teacher near Henderson, how might my bosses react? See me as raising the alarm again? Crying wolf? It didn't help that wetlands, prime habitat for frogs, were not a priority at the agency. Not valued like lakes and rivers.

I wanted to get to Cindy's pond immediately, before the frogs dispersed to feed on bugs in upland areas. In 1993 we hadn't been notified about the Granite Falls frogs until mid-September. We'd found a few with deformities, but by then most frogs had scattered, probably to forage and prepare for winter.

But this time we had a chance. It was August. Frogs were at the pond, Cindy said.

That week I was committed to work with the US Environmental Protection Agency (EPA) to produce a film about the value of protecting wetlands. The EPA, which funded all of my work on wetlands at the Minnesota Pollution Control Agency (MPCA), wanted to encourage state pollution agencies to better protect wetlands water quality. My supervisor, a by-the-books kind of guy with a background in civil engineering, had tried, unsuccessfully, to stop my participation in the filming, saying our agency's leadership "had not yet agreed on its policies toward wetlands."

That government agencies would work to protect wetlands water quality was not widely accepted in the early 1990s. In the early 1980s, the Reagan administration and a pro-development, Republican Senate stood against environmentalists who hoped to strengthen the weak protections given to wetlands at the time. But in 1985, perhaps as a cost-saving measure, federal farm policy changed dramatically. Farmers could no longer collect government subsidies for crops grown on land they wanted to convert from wetlands to cropland. And soon after, the tax incentives that had previously encouraged wetlands conversions were also eliminated (Vileisis 1997).

But wetlands losses and degradation continued. Protecting the water quality and natural values of the nation's remaining wetlands received little attention. By 1993 the EPA began to encourage scientists, myself included, to wade in and develop methods for measuring the biological health of wetlands. Biological monitoring could assess the degree of pollution in wetlands, especially when chemical pollutants in wetlands were not monitored. In the early 1990s no one at state pollution control agencies measured the actual water quality of wetlands, let alone their biological integrity. Water quality of rivers, lakes, groundwater—yes. Wetlands? Not likely. At best, wetlands served as catchalls, places to run polluted water so it didn't harm other, more highly favored surface waters. Regulating wetlands was politically unpopular among farmers and developers alike; it didn't help that many are located on privately owned land.

Just before Cindy's life-changing phone call, I had been out sampling dozens of ponds to fulfill EPA's charge to design biological monitoring methods to assess wetlands water quality. If their water is clean, small vegetated ponds nurture a fascinating diversity of creatures, not just the aquatic invertebrates I sought, but also long-legged wading birds, secretive marsh wrens, nesting waterfowl, colorful flowering aquatic plants, spiky rushes, grasslike sedges, salamanders, and reproducing frogs.

As dawn was breaking one morning, I visited a shallow, clear-water wetland located in central Minnesota in a federal waterfowl production area named Kenna to sample the aquatic life. I waded past wild blue iris near shore and into the water. I stood quietly, barely able to see the emergent stalks of delicate white flowers and low-lying plants with wide, arrow-shaped leaves.

Light from the rosy dawn colored patches of open water while wisps of morning mist drifted upward. A few male frogs still called to attract mates, their now diminished chorus having begun at twilight the evening before.

From the road, this wetland looked unimpressive. But I found it harbored a surprising variety of tiny species uniquely suited for life in a pond: shiny dark beetles, whose females laid eggs on the backs of the males; unpresuming mayflies, destined to emerge from the water on gossamer wings, mate, and die; variously shaped snails, little crustaceans, and miniature clams; immature dragonflies, whose lightening-fast jaws trap mosquito larvae and other underwater bugs. An invisible, sensitive world. A safe environment for water-loving creatures like frogs. The diversity in that wetland evoked the joy I had experienced outdoors in various places in New England where I grew up: rocky streams, home to native brook trout, in the mountains of Vermont; weedy fringes of lakes dense with tiny black tadpoles; beds of sea grass populated with horseshoe crabs and scallops in the Atlantic.

Not all wetlands matched this little gem. One polluted wetland I waded into looked more like a war zone. Its muck pulled on my booted feet, its emergent clumps of wetlands plants lay keeled over, dying in the murky water. Years of erratically fluctuating, contaminated storm water that flowed from the ever-expanding urban developments had decimated this wetland's invertebrate life. Only the most pollution-tolerant species survived the onslaught. This pond needed life support.

"My gut instinct is to come tomorrow, before the frogs move away from the pond," I said to Cindy, explaining my tight schedule for the EPA film shoot. "I'll ask Joel, our student worker, if he could meet you there. What's a good time for you?"

"Tomorrow at noon," she said gratefully. I asked her to describe the pond, embarking on what would become a long relationship of shared communications.

"The pond is on the Ney [pronounced "Ny"] wildlife area, one of my favorite places to take students," she said. From Cindy I learned the Ney family had farmed the land for over a century and had recently donated it to the county. The bean-shaped pond had been created on their farm several years earlier by excavating a wet spot. Sometimes its water spills over at one end and flows down a ravine to join a small creek that runs into the Minnesota River down below.

"Sounds like the river can't possibly flood the pond," I said, remembering the floodplain wetland in Granite Falls where deformed frogs were seen in the owners' backyard, which had been flooded by the river. In spring, water from the nearby Minnesota River, the state's most polluted, has historically overflowed its banks south of Granite Falls, in some years flooding the owners' property and threatening their home.

Floodwaters were already on my list of sources of possible harm to wetlands and frogs. In 1993 an amphibian biologist had sent me an unidentified newspaper article headlined "Mighty Mississippi Floods Cause Mutant Frogs." Deformed frogs had been discovered in river-flooded wetlands in Iowa close to a chemical plant just north of Muscatine. The article pictured a frog with a leg growing out of its back. Other frogs had extra legs or branched legs. A local biologist was quoted, saying, "It's very unlikely it's inherited; it's more likely a response to an unnatural condition."

Cindy Reinitz taught at the Minnesota New Country School, a new charter school for grades seven to twelve then located in abandoned store fronts in the small town of Le Sueur, home of a vegetable processing plant. Its signature Jolly Green Giant sign stood high on the bluff above town. "The students will want to gather information about the frogs and do water tests," Cindy said. "Already they are posting their observations about the frogs on the Internet." She concluded our conversation with directions to the pond.

Slowly I put the receiver down and turned to my colleague, Mark Gernes (pronounced Gur'ness). I took a deep breath. "Here we go again," I said. "Only this time *lots more* deformed frogs."

Mark and I worked closely together to convince our agency to value clean water for wetlands. With Mark's expertise in wetlands plant communities and mine in aquatic invertebrates, we had teamed up to research ways to monitor wetlands water quality by examining the invertebrates and plants they supported. We wanted to let the biology tell us if a pond was healthy or sick. Luckily for us, the federal EPA agreed and for years funded our wetlands research.

Tall and fit, Mark could easily leg it through all sorts of wetlands and regularly work long days in the field. His easy-going and humble demeanor belied the inner drive and unflagging energy that he poured into our work. His loosely waved brown hair and close-cropped mustache framed a face that reflected a steady kindness, a face that, unlike mine, rarely frowned. Soft-spoken and always humble, Mark would not take credit for what he'd accomplished and had little appetite for confrontation. A good twenty-three years younger, he had no idea how much his quiet commitment to our work helped sustain me. While I tended to complain and fight the system, Mark would hold a steady course.

Mark's kinship with nature developed during youthful times as he hunted with his father near natural wetlands and roamed through remnants of native prairie in central Minnesota, where he grew up. Once, on a church retreat, Mark and other teenagers were told to go off alone for a few hours of personal visioning time. After wading in a stream, he piled up some rocks and stepped back to think. Mark told me he decided then to devote his life to protecting rivers and the environment. The stacked rocks, he said, "seemed like a good foundation for what I didn't know."

Mark earned his undergraduate and master's degrees in biology, working summers as a naturalist in state parks and doing fieldwork for the Nature Conservancy, documenting plants in a wet prairie. He did research on prairie plants at Wes and Dana Jackson's pioneering Land Institute in Kansas. By the time I met Mark, he was so knowledgeable he could quietly identify just about any kind of plant he saw.

Mark came to MPCA the same year that I did, in 1989. At first he worked in a classified—that is, permanent—position in the program that oversaw leaking underground storage tanks. Three years later, he left that permanent job to join the two of us who worked on biology and water pollution. I was quite impressed that this young father with two small children would take such a risk and leave his more secure position.

Our work on wetlands evolved through the 1990s. At first EPA funds sustained us for short periods, a few months at best. Then we landed larger grants, often covering two years, to support our work to research biological indictors for wetlands. Getting the grants was always a pressure cooker for us. Unlike other MPCA staff whose work was also supported by the EPA, Mark and I had to cover our salaries and fringe benefits. Absent those grants, our work would expire. We'd be out of a job.

Our backgrounds differed. I grew up in Massachusetts and had loved fishing in mountain streams in Vermont, swimming in lakes, or riding ocean waves. Early on I became entranced with water. But I knew little about wetlands other than the beaver ponds we crept up to while trout fishing in Vermont, or the boggy areas we dodged when hiking along streams. Before landing my job at the MPCA, I had joined with EPA scientists for a summer project to research some wetlands that had been experimentally treated with pesticides to test their effects in a natural environment. Seeing firsthand the devastation chemicals wrought on pond invertebrates solidified my future path: to work to protect these fragile, watery habitats and their pollution-sensitive organisms.

Cindy's call marked the beginning of a long and challenging journey into unexplored territories, a trip that would take me deep inside the vulnerable world of frogs, the tangled forest of politics and government bureaucracy, and the complexities of environmental science. It would be a time of personal change when I would risk forming a new relationship, a time when I would be humbled on a regular basis by what I did not know, a time when I would be continually challenged to keep learning, and learn fast. Ahead lay obstacles that would force me to draw deeply on the inner resources I depended upon to keep working for what I believed in.

I asked my summer assistant, Joel Chirhart, if he could go to the Ney Pond Tuesday to meet the teacher, Cindy Reinitz, and examine the frogs. He agreed, glad to escape the tedium of identifying and counting invertebrates at the microscope in the MPCA's windowless basement lab. Easy-going and flexible, Joel liked being out with people in the field. He would be a good person to connect with Cindy and her students.

The next day, August 15, Joel drove south from St. Paul to the Ney Pond, his state-owned vehicle loaded with nets and supplies that included a cooler with dry ice because I had asked him to collect and freeze frogs for tissue analysis later.

At a wetlands workshop I had just attended in Montana, scientist Dennis Lemly described how deformities in birds and fish had developed at a marsh created in the California desert during the early 1980s. Excess irrigation water from crop fields was channeled into a new wildlife refuge named Kesterson. Soon waterfowl and other wildlife began reproducing in the watery habitat.

Then disaster struck: ducklings hatched out with corkscrew beaks and misshapen or missing wings and legs. One had its brain held in a sac outside the skull, and others had bulging or missing eyes. Fish and frogs also grew with deformities. Signs were posted to warn people not to fish or hunt. Noise-producing cannons blasted sounds to drive away birds from the toxic marsh, to no avail.

The culprit in this mystery was the metal selenium, leached from agricultural soils by the irrigation water that collected in perforated pipes underground and then flowed into a canal that drained into Kesterson. Eventually, the contaminated drainage water had to be diverted and contained in small farm ponds so the desert marsh could dry down. Then the selenium-poisoned soil was scraped off and hauled to a hazardous waste disposal site. Had this not been done, the canal's toxic water would have been conducted to the San Francisco bay.

I had told Dennis Lemly about the deformed frogs in Granite Falls.

"Finding the selenium in the tissues of the deformed birds gave us a major clue to the cause," he said. "You will need to examine the frog tissues for contaminants." Lemly had also described some of the political roadblocks that confronted federal fish and wildlife researchers. One agency (the federal Bureau of Reclamation) had tried to repress the scientist's results, not wanting the public to learn of the selenium toxicity. The dramatic story was chronicled by journalist Tom Harris in his book *Death in the Marsh* (1991).

The eventual outcome of the 1993 meeting that summer in Montana was a book on ecotoxicology and risk assessment for wetlands (Lewis et al. 1999) written by several working groups of scientists in attendance both from government and industry. I had mixed feelings then about the use of "risk assessment" as a scientific method for making decisions about responsibility for specific pollutants in the environment. Certainly one should take into account the degree of damage chemicals can cause to human or environmental health when imposing regulations. But could a risk assessment approach be misused as a way to reduce or minimize a polluter's responsibility for cleanup?

While at the meeting I talked quietly with one of the biologists about news coming out of Australia, where six species of tropical rainforest frogs had disappeared and populations of other frogs were shrinking. Early examination of Australian frogs revealed a type of skin fungus later shown to mortally infect frogs. Australian scientists warned that up to 26 of the nation's 202 species of frogs were in decline. In Costa Rica's Monte Verde cloud forest, populations of the brilliantly colored golden toad had plummeted in the late 1980s, I'd later read, along with other species of amphibians in that area. Biologists in our country were also seeing warning signs. Several species of frogs were in sharp decline in California, Colorado, and Oregon. Worry spread. What was happening to the frogs?

I headed home from work late after Cindy's call. I wondered where this new report of deformed frogs might lead us. I had been hoping we'd have a slower pace at work after our intense summer sampling wetlands in central Minnesota. Plus, I was in a new and intriguing relationship and wanted time to explore it. Divorced for years, I treasured my independence. But recently I'd met Verlyn Smith, a retired, progressive pastor and college teacher, and we'd begun to talk cautiously about marrying.

Late on Tuesday, August 15, Mark and I were on the road in central Minnesota for the EPA film shoot when our clunky portable phone rang.

It was Joel, sounding shaken, an urgent tone in his voice.

"Judy, you *have* got to go down there. Those frogs are awful," he said. "Lots of them are grossly malformed. Maybe forty of a hundred I looked at had deformities. It's unbelievable."

Mark, who was driving, looked over with a sharp eye.

"This sounds much worse than Granite Falls," I said. I asked Joel if he could go down again Friday afternoon after the EPA film crew left.

"Sure," he replied. "But it's not going to be pleasant."

On Friday, having said farewells to the EPA film crew, Joel and I drove south from St. Paul to the Ney Pond. After a couple of hours, we turned off the highway and bounced along a dirt access road, then swung left through a cornfield. Ahead I could see a pond surrounded by corn and soybean fields. I knew Cindy Reinitz couldn't be there that day, but others would be.

To our right sat small cattle sheds and an old farmhouse, Don Ney's place. Near the pond, in an open, grassy area, stood clusters of kids and adults. Some looked excited; others appeared troubled as they watched us roll in. Beside the few parked vehicles lay buckets, nets, day packs, and loose boots scattered around on the ground. I would remember the scene forever.

After introductions, Joel and I sat down on the back fender of a van. Eager students and concerned parents stepped close. They knew what I was about to see.

I reached into the five gallon pail, its bottom crowded with frogs the students had collected. I grabbed one, and then another. The first frog had one rear leg completely missing—like a total amputation except the skin looked normal. The next had a stumpy leg, half the normal length. It swung around uselessly. "This is awful," I said quietly, seeing a couple of parents nod their heads sympathetically.

Each frog squirmed in my hand as I gently turned it over to check its body. The flesh of their white underbellies felt clammy and cool in spite of the summer heat that enveloped us. I flashed back to my college days, to the glistening beauty of the internal organs—the liver, heart, intestines, lungs—of the freshly killed frogs that we dissected in zoology lab. What might these frogs look like inside? I wondered.

"Some in there have extra legs," a student piped up. Joel rummaged in the bucket and lifted out a little frog. An emaciated, useless-looking third leg projected out over the normal leg on one side. "How weird," Joel murmured.

Feeling my stomach churn, I looked up at the expectant faces. I glanced at the pond. Its water was calm, its shoreline fringed with tall grasses. So innocent looking, so rural, a place for kids to learn about birds, bugs, plants, and pond life—not handle deformed animals. Upslope, on the far side, a lone bur oak tree stood like a sentinel. Above the venerable tree, at the crest of the hill, lay farmed cropland. How could this pond have brought forth such monstrous and pathetic creatures from its waters?

It felt like some kind of ghastly science-fiction film.

Was this something natural? Caused by humans? I couldn't think. Could it be caused by parasites that emerged from intermediate stages of snails living in the fringes of the pond, parasites I'd read were suspected of causing multiple-limbed frogs in California? In that case, the people who lived around there couldn't possibly be harmed. Or could it be one of several pesticides used on the adjacent crops, or harmful rays from the sun? Or something else? Disease? Damage from predators like herons or voracious water beetles nipping at limbs? Why now, why here, in this innocent-looking pond? And how could this be sorted out when we knew so little about what could actually cause such aberrations?

The weather was uncomfortable—well into the nineties. Joel and I had already experienced many steamy days sampling wetlands for invertebrates in our overheating, chest-high waders. But this time we didn't notice it, we were so completely immersed in our task: recording deformities in the frogs the students had collected for us and capturing others with our long-handled nets. Not a random, scientific sample. What mattered was the astonishing number of frogs with malformed bodies.

By late afternoon, people were headed home for supper. I asked Joel if he'd be okay staying late to help photograph the frogs. I had an urgent need to document the deformed frogs on film, so the public would believe us and pay attention. "This time, I hope to God management understands there's a real problem in a wetland," I said.

It was getting late and still near ninety degrees when Joel and I finally settled down to shoot pictures with the three cameras I'd brought along, just in case one camera malfunctioned. I shot photos of many frogs held gently but firmly in Joel's hand, still dirty with pond mud. The best images, taken with my old Olympus camera and macro lens, later went worldwide.

Some of the deformed frogs that had extra or branched legs were sensationalized in the media (they *were* bizarre-looking), making people think these types of abnormalities predominated. But they didn't. Already we could tell from our collections that the majority of deformities were either partial or missing rear legs.

In the early evening shade, Joel and I sweated heavily as we measured the body length of each frog from the tip of the snout to the vent at the base of the hind legs, per instructions from Dave Hoppe, the biologist from the

University of Minnesota at Morris, who'd worked with us on the Granite Falls frogs. Our measurements confirmed what Cindy had told me: the deformed frogs were newly developed, not adults. Below are some of our crude descriptions recorded that day.

- leg twisted, foot back
- a 3rd leg extends, as if no muscle
- left leg all twisted, close to body, has extra, thin stump
- severely contorted leg base, can't see leg, jutting (left side)
- left leg missing, no stump even
- 2 legs at left leg plus a right leg = 3 legs, abnormal one is anterior to normal
- right leg thin, skin pulls like it's too tight, probably can't extend fully (as if webbed across the thigh)
- right eye missing

Of 53 frogs we examined that day, 60 percent of them had malformations. A week later, we collected 137 frogs at the Ney Pond. Of these, 105 (more than three-quarters) had deformities. Such numbers startled us. No wonder the teacher and her students were alarmed.

While Joel and I worked late, Sarah Malchow, the reporter and co-owner of the local Henderson newspaper, reappeared. She had been there earlier in the day and now returned toting water, saltines, and fresh tomatoes from her garden. In our rush Joel and I had both forgotten to bring anything to drink. "An angel," I thought. Water and saltines. Delicious.

Then, animated, she launched a string of questions. "What could cause this? What can be done about it? Has this happened before?" I snapped back to reality and tried to answer as best I could. She took notes.

Sarah's article ran the next week in the *Henderson Independent News* and was one of the first published written piece about the deformed frogs (Malchow 1995). She summarized our plans to analyze frog tissues and didn't suggest that answers would come anytime soon.

Earlier that afternoon veteran TV environment reporter Ken Speake had been at the pond. I welcomed his coverage, knowing he'd be fair and thorough. He'd been out with Mark and me in Granite Falls, and as then, Ken shouldered his heavy video camera and waded right into the water with us, talking and recording as we caught frogs. Ken interviewed Joel, me, and the students, and that night his piece ran on the six and ten o'clock news, the first televised coverage of Minnesota's deformed frogs.

Verlyn and I had made plans for supper that evening, but knowing I needed to stay late at the pond I'd had to cancel. Finally home, I called him after nine and described the awful frogs I had seen that day. He listened quietly. He didn't have to be a biologist to know that something ominous had happened.

"Watch the ten o'clock news, then you'll see what I'm talking about," I said.

At ten I ate supper and watched Ken Speake's piece. It opened with Joel in a tee shirt and hip boots wading in the pond. A frog swam erratically in the water near some weeds. Joel grabbed it and examined it closely, then said, "Omigosh, this one's deformed in both legs; I think this is the first one with deformities in both legs."

Speake filmed me explaining that frogs are good indicators of the health of a wetland or a river system, because as they grow from water-dwelling tadpoles to land-dwelling frogs, they can easily be damaged in many ways. I was shown holding up a frog with a weird, conical projection sticking out from its groin, then another frog. "Oh, look at this. That is *really* bizarre," I say. "It has extra legs!"

Ken interjected that half of the frogs collected by Cindy Reinitz's students had some kind of deformity, and I commented that it was very unusual to see this many deformed frogs.

"The frogs will go to a laboratory to be analyzed, and when they find out what might have caused the deformities, they'll try to figure out what to do about it. This is Ken Speake, KARE 11 News, at Henderson." Then the station's news anchor concluded in a solemn voice: "There are no immediate threats to residents. The affected frogs are not eaten by people; they are not in the human food chain."

Soon after, teenager Nate Sellner showed compassion for the frogs themselves in an interview on the radio. He blamed agricultural runoff and said he'd "like to find the cause, if only to stop the frogs' suffering. There's one that we found that only had one leg, and it kept jumping. It would land on its back. They'll be more likely to be eaten by predators. This is big and let's just find out what this is," he said. When Joel and I had collected frogs that Friday, I'd seen this—the abortive leap of a one-legged frog that landed on its back, its white belly exposed, its limbs flailing.

Were we at the beginning of something big, or could this pond be like the wetland near Granite Falls? Was this just one more isolated report of troublesome frogs from a site that might not spawn deformed frogs in years ahead? Another fluke, another peril of the moment, something to observe and then try to forget?

In early September news of the gruesome frogs and our nascent frog investigation went statewide when the Minneapolis *Star Tribune* ran an article on the front page titled "Deformed Frogs Prompt Investigation" (Rebuffoni 1995). Following that, the national media swooped in: something extraordinary, they reported, was happening to Minnesota's frogs.

At the time the article appeared, I was up north for a much-needed break from the intensity of my job while Verlyn went to South Dakota to help his aging mother. For my week-long retreat in a cabin by Lake Superior, I wanted to read and prepare talks that I would give to encourage my fellow church members to embrace care of the environment as a religious issue and a responsibility. The church already had a full plate of concerns related to human suffering. Could it make room for the rest of creation as well?

Outside of work I had talked to an assembly of various church people and members of a coalition of groups organized to clean up the Minnesota River. I explained my work and how my faith helped drive my feelings about reducing pollution to protect both humans and the environment, including vulnerable species like frogs. In the discussion period, one attendee, who believed the pollution in the river was natural and not caused by humans, handed the moderator a statement that "the Minnesota Pollution Control Agency is a whitened sepulcher"—a biblical reference to hypocritical leaders who appeared pure white like mausoleums on the outside but on the inside were full of decay and bones. As that gathering ended, I had asked someone to escort me to my car, unsure what this man's attitude might be toward MPCA employees.

Later I learned of an organization called the River Warren Research Committee, or RWRC. Strongly antigovernment, RWRC accused government staff of spreading false propaganda about pollution in the Minnesota River in order to gain control over people's land and lives. They decried government regulations, advocated reducing its size, and promoted individual liberties. These attitudes still simmer in Minnesota.

The River Warren group promotes the notion that the soils muddying the Minnesota River do not erode from the farmland that comprises most of its watershed. Instead, they claim, the soil in the river is washed there from the historic, five-mile-wide basin carved out ten thousand years ago by the River Warren when it gushed from glacial Lake Agassiz. That ancient lake was a large body of glacial meltwater located in northwestern Minnesota that broke open on its southern end, unleashing a huge flow that carved out a new river basin through which the Minnesota runs today. Farming practices had little or nothing to do with the muddy condition of the river, in RWRC's view;

drainage of wetlands did not exacerbate the erosion of soils from tilled cropland; restoring wetlands will not make any difference either in flood control or in soil erosion.

These beliefs stand against the established science that most of the mud in the river derives from the erosion of exposed soils coming from the farmland that covers 90 percent of the enormous watershed of the Minnesota River. In addition, most of the wetlands (again, 90 percent) have been drained in the southwestern area of the state. Subsurface drain tiles buried below these former wetlands, now cropped, convey silt and farm chemicals to county ditches that flow to the river.

While on my retreat I walked along Lake Superior's stony beach and listened to the clinking sounds of stones rolling in the surf, or strolled through the woods and heard tree frogs calling. I couldn't help but reflect on my relationship with Verlyn, about some differences that had arisen. Yet we agreed on many issues and on the importance of acting out our faith in our lives.

When I returned to work after my vacation I faced an onslaught of media interest triggered by the recent newspaper article on the frogs. Radio stations and TV networks called me for interviews, to the point that our division director kidded, "Pretty soon we will see Mark and Judy in the tabloids." Eventually we were (Ziegler 1996).

Gail Thovson, a middle school science teacher, called from Litchfield, an older town a couple of hours' drive west of the Twin Cities. "I read the *Star Tribune* article about the frogs to my science class," she said, explaining how she'd projected the article, with its photos of three deformed frogs, on a screen and opened up a discussion with her students. She'd asked them what they thought had happened. What might the future bring? Thovson told me later she liked to foster investigative learning and let her students form their own ideas and conclusions. The students could hardly believe something that awful could happen to frogs, she said.

The next day, a boy, known to be a class cutup, came into school toting a bucket of severely deformed frogs. Suspecting he was up to some mischief, the teacher thought to herself, "I'll bet a nickel he took off the legs."

"Of all people, it would be this kid who would find them," she later told me. But she and the class saw no signs of injury. This was no prank. The deformities were real.

Having live deformed frogs to care for in the classroom galvanized her students into action; they began to ask a lot more questions. They brought insects and worms to feed the helpless creatures, one of which was missing both rear legs, Thovson said. When the students tried to feed a one-legged frog, it would fall over on its back and had a hard time rolling over to right itself.

Someone brought a deformed baby bird to class, a dead grackle with three legs. Seeing the bizarre bird both intrigued and spooked the students. Could other animals be harmed? Could they themselves be harmed?

Thovson contacted Audre Kramer, the homeowner on whose land the boy had found the awful frogs. Audre was willing to talk with me. "Most of our frogs are deformed, and other people around here are seeing them also," Audre said when I called. She'd found the frogs around a pond that she and her husband had excavated in their yard two years before. "We wanted the pond for wildlife, like frogs and ducks, which we love," she said. I arranged for a visit.

Thovson also suggested I talk with an older man who'd caught some deformed frogs near the golf course in Litchfield. I did. "We was getting 'em to fish for northerns, you know," he said. "We like the frisky ones. But some had bad legs like a canoe paddle with no toes, or just half a leg on both bottom legs. I felt sorry for the little rascals," he told me.

The water where he caught the frogs flows off the golf course into a ditch that drains to Jewett Creek. I wondered what chemicals the golf course used. "I've been gettin' frogs there for fishin' for twenty years. They've been real scarce lately, but I've always found 'em near the golf course. This is the first year ever that I've seen these deformities. I'm sixty-seven years old, and there ain't too many things that get me excited any more," he said.

At the end of September, I drove west to Litchfield with our field assistant, Jon Haferman, and Bob McKinnell, a University of Minnesota professor in the Department of Genetics and Cell Biology. Bob's main research centered on cancer biology and certain viral-caused tumors, but he also had an enormous interest in leopard frogs. For decades both he and fellow zoologist Dave Hoppe had surveyed frogs across Minnesota and in the Dakotas to study the genetic distribution of the skin color patterns.

As we drove, Bob and I discussed the life of northern leopard frogs, which, like Hoppe, he knew inside and out. These frogs have a vagabond but purposeful life: they survive the winter in deep water. They migrate overland in spring to breed and lay their eggs in shallow, fishless ponds. In summer they roam away from wetlands into vegetated areas to hunt for bugs and worms, and by fall most leopard frogs find their way to deeper water and hunker down for a long winter. They have to. No wonder people call frogs the canaries of our environment, trite as it sounds. Frogs are exposed to pollution both on land and in water.

I briefly updated Bob on the work that Mark and I (and biologists in other states) were doing to develop ways to analyze the health of wetlands. But this was not the time to display my belief that, long term, seeing declines in the numbers of pollution-sensitive species of aquatic invertebrates and water-dependent plants might ultimately prove a better measure of wetlands health than what's happening to frogs. Or that my colleagues in our biological monitoring group effectively used measures of the fish and invertebrate communities as the "canaries" of river health. Instead, I wanted to learn all I could about frogs from McKinnell.

In the rain, we drove north from Litchfield past farm fields, then turned onto a gravel drive to Audre's home, which sat surrounded by an expanse of manicured lawn that sloped to the dark-flowing Crow River, a good-size tributary to the Mississippi. Audre came out to greet us from her open garage, where I could see folding chairs set up facing an aquarium on the cement floor. Together we walked a short distance across the wet grass to the small, fifty-foot-wide pond created in the sloping yard.

Audre explained they'd hired a contractor who dug a trench downslope of the pond and put in a clay berm underground to retain water.

The small pond looked completely artificial to me. Its gravelly sides were way too steep to hold rooted vegetation; the lawn was neatly mowed right to the edge at the top. Too sterile; no bugs for frogs here, I thought. "Sometimes the water turns green," Audre said. Floating in the pond was a small raft roped to a pulley system, a platform on which they put corn for ducks and geese. Little wonder the pond greened up—the grain and the duck feces would fertilize the water and boost algae growth.

In a misty rain we netted northern leopard frogs all around the pond's perimeter, then settled ourselves on the chairs set inside Audre's garage to examine them. The barren habitat of the pond and the severity of the deformities had taken a toll on the young, newly metamorphosed frogs. Dwarfed and emaciated, their bellies lay flat and flaccid. They were, understandably, lethargic. Many were missing a leg, or had tiny stumps where a leg should be. How could they possibly capture insects, even if any bugs survived in this pond's unnaturally tidy habitat? Frogs needed to jump quickly and zing out their sticky tongues fast enough to trap an insect on the fly.

Two frogs had no rear legs. They struggled to pull themselves forward with their front legs in the shallow water, a heart-wrenching sight. I found myself looking away from the most severely crippled frogs. Years later I read of a reporter who accompanied a soldier after Hurricane Katrina while he boated around to tag bodies that still floated in the water in New Orleans. She did not want to look at the swollen corpses, but the soldier insisted, saying, "Look at him, *you look at him,* so you can tell people how bad this really was." In retrospect I realize we'd felt like that reporter. We *had* to look at the frogs and we *had* to tell people how truly awful this was.

We worked in stunned silence, only speaking in solemn tones to record what we saw: "Thirty-three mm, only a nub for a right leg," one of us said, using our term for a very short stump. The litany went on: 31 mm right nub; 33 mm left nub; 33 mm both legs nub; 32 mm no right leg; 34 mm left leg shortened and weird; 33 mm no left leg, almost no right. Of ninety-one frogs, only *two* were normal. Our assistant, Jon, wrote, "Freed a normal frog—yay!" in the field notes.

Suddenly, Bob McKinnell broke the silence: "These look like *thalidomide* frogs!"

I remembered the thalidomide scare—the drug given to many pregnant women to prevent morning sickness, the drug that caused nearly ten thousand babies to be born with shrunken, defective limbs. An image flashed through my mind: those photos I'd seen once of a crying baby with no arms, its misshapen feet attached to the pelvis, of another child with a deformed, partial hand at the end of a tiny stump. There was that woman I'd seen once in Boston: she had a hooklike appendage that shot out from her shoulder to scratch her head, then disappeared under her cape.

Marketing of thalidomide in the United States was blocked in the early 1960s, the day before it would have been approved for sale by the Food and Drug Administration. Shortly after, in 1962, Dr. Frances Kelsey received the President's Award for Distinguished Federal Civilian Service from President Kennedy for her action that saved thousands of US infants from major birth defects. Dr. Kelsey's decision showed the positive role that government could play in protecting human health. In the 1960s the thalidomide disaster raised alarms: shouldn't government be testing drugs to see if they cause birth defects before they are given to pregnant women?

And what about Vietnam? Some have suggested that birth defects in babies born during and after the war there resulted from our military's extensive operation spraying the herbicide Agent Orange over huge areas of that country in the late 1960s. This herbicide was contaminated with dioxin, a developmentally toxic chemical by-product created during the manufacturing process.

As I handled more frogs with stumps of limbs, I kept wondering what could have caused such abnormalities. I wondered how many chemicals used in rural areas had been properly tested before they were released. Were they screened in advance to determine whether they might cause birth defects? Or was it something natural, like immature parasites that interfered with normal limb development or perhaps predators like fish that might nibble the developing legs?

It was remarkable that any of the frogs we examined that rainy day could move at all with such tiny distorted stumps. It seemed weird to see a quarter-inch stump wiggling around. “It must have some muscle attached somewhere,” I said, my skin crawling. Other stumps just stuck out, immobile, as if made only of skin and cartilage. Could any of these frogs survive the winter? Could they make it down the hill to the Crow River, where they'd need to spend the winter underwater?

All the deformities in the frogs from Audre's pond were in the rear legs: missing limbs, little stumps, or twisted limbs. We saw no extra legs, no skin webbing, no missing eyes. Why was this site so different from the Ney Pond, whose frogs had a wider array of deformities?

We debated whether digging a cavity in the yard to create a pond for wildlife might have played a role. Did the excavation dredge up old soils and release buried pollutants? Audre said there used to be some kind of dump right near the pond over twenty years ago, but she doubted the farmer who worked the land had used much pesticide. “He was too cheap and not smart enough,” she said. Or, we wondered, did this pond collect drainage water from the farm fields upslope, across the road?

The Ney Pond had been created in 1989 by bulldozing a wet area. Soil was pushed up to make two tiny “wildlife” islands, presumably to create sites for nesting Canada geese. The pond receives water drained from Don Ney's and a neighbor's crop fields for some distance through four underground agricultural tile lines, the perforated pipes positioned under fields to drain away excess water. What role, if any, did pond excavation play in the frog deformities? Or farmland runoff?

Audre Kramer was recovering from ovarian cancer, and she told us about the high number of other people with cancer who lived close by. “I've been mapping them. I believe there's more than normal here,” she said, as she unfolded a map on which she'd made numerous marks to locate the homes of people with cancer. The rural area where Audre lived was sparsely populated. She'd found out about the other cancer patients while trading stories with them in the waiting room at the clinic. She knew where they lived and what kinds of cancer they'd had.

Later, I asked an MPCA staff hydrologist if he could drive out and look at Audre's land, to see if he could determine the pond's water source. When he returned to the office, he told me the pond was most likely fed by groundwater that came from the agricultural land upslope. "Who knows what the excavation may have brought up," he added. The cancers located on Audre's map were all in a small watershed, he said, suggesting that the nearby farms could have the same groundwater (and well water). Studies have found that many private rural wells used by farm families contain pesticides. Was there any link to Audre's situation?

I was concerned for Audre. What did this apparent concentration of cancers in an area that had deformed frogs mean about the local environment? Could there be something really toxic there, or was it just coincidental? Was this a "normal" rate of cancer? No wonder she was alarmed.

Back at work, I called the state's Department of Health for advice about Audre's cancer records and was directed to Al Bender, head of Chronic Disease and Epidemiology. I described to him Audre's map, the number of cancers she had recorded in the surrounding local area, and what our hydrologist had said about the area being in a small watershed with shared groundwater. Bender spoke authoritatively, like a man certain of what he knows. "You know, most of the time, when people think they are in a cancer cluster, they simply need counseling," he said.

Stunned by his comment, I restated Audre's records and her concerns. It seemed like a lot of cancers were clustered in such a small area fed by the same groundwater, I said. He responded, "Really, when you get down to it, there's been no definite link between an environmental agent and cancer." I was astonished by this. Later in an article in the *New Yorker* about "the cancer cluster myth," Bender was quoted as saying, "The reality is that they're [investigations of cancer clusters] an absolute, total, and complete waste of taxpayer dollars" (Guwande 1999). He expressed pride that Minnesota had such an effective public-response system, that it had not needed to conduct a cancer investigation in three years. What system, I wondered. Counseling?

Bender directed me to Health's cancer epidemiologist, who told me that cancer rates might be elevated in the rural county where Audre lived and sent me a report on cancer rates in counties in Minnesota. That was it. No offer to examine Audre's map or have someone in his department call her. Later, I apologized to Audre because no one from Health had followed up her concern or contacted her. Absent an investigation by health experts, nothing could be concluded about the numerous cancers she had recorded on her map.

This attitude from our state's Health Department puzzled me. I realized it had been extremely difficult to link environmental agents to specific human cancers, but it had been done. Lung cancer and cigarettes, for instance.

Workers exposed regularly to asbestos had developed a rare type of lung cancer. Those involved in the manufacture of polyvinylchloride pipes could get liver cancer. Cancers had been pinned to particular drugs, like diethyl stilbestrol (DES), for example. DES, widely prescribed in the United States from the late 1940s to late 1960s to prevent miscarriages and other reproductive problems, caused an extremely unusual form of cervical cancer that developed in daughters of mothers who had taken it during early pregnancy (Colborn et al. 1996). Later, effects were also found in DES-exposed sons.

I was very aware of the challenges scientists had in pinning down trichloroethylene (TCE) as the cause of the cluster of childhood leukemias in Woburn, Massachusetts, close to my home town. Wells for public drinking water were contaminated with the toxin, but in people's homes the pollutant was measured at safe levels in their tap water. Not until the EPA discovered that the greatest amount of TCE exposure occurred in showers, where children inhaled the chemical (which had been volatilized by the warm water) was TCE considered the most likely cause of the children's leukemias—a story told in the book *A Civil Action,* by Jonathan Harr (1995).

Audre lived downslope of agricultural land. Could she and her rural neighbors have been exposed to pesticides that might have triggered their cancers? During the 1990s, many pesticides had been shown to cause cancer in animal tests—one study found that 24 of 240 pesticides tested could cause thyroid cancers, for instance (Hurley et al. 1998). Associations were found between pesticides and certain cancers in epidemiological studies of people who were exposed to pesticides, such as farmers, golf course workers, and pesticide applicators (Dich et al. 1997).

As part of its regulatory process, the EPA reviews scientific studies for evidence that a particular chemical might cause cancer, using carefully designed criteria for reviewing scientific reports. They rate the probability of cancer causation and decide whether a chemical is definitely a carcinogen, a likely carcinogen, or not likely (US EPA 2009a, b; Dich et al. 1997; Hurley et al. 1998). Many pesticides are rated by the EPA as likely or probable human carcinogens (US EPA 2009a, b). For instance, the insecticides permethrin, resmethrin, and carbaryl are classified as "a likely human carcinogen," the herbicide acetochlor and the insecticide malathion have "suggestive" cancer-causing potential. Maneb, a fungicide used on potatoes and tomatoes (and later found in one deformed frog pond) is listed by the EPA as a "probable" carcinogen. The herbicides atrazine and glyphosate are "not likely" to cause cancer. Based on their assessments and other toxicities, the EPA can recommend safe levels of the chemicals of concern or it can restrict or ban a chemical's sale altogether, a process that can take years. If the sale of a chemical is blocked, farmers are usually allowed to use up their existing stocks.

In the fall of 1995, as the media reported the students' discovery of the frogs, several people called me about deformed frogs they'd seen in different places: by the Rainy River in northern Minnesota; in Ontario; by a farm pond in Texas. A woman who called from Ohio said, "Two thirds of the toads on our farm are deformed. They have missing legs and extra legs." She had frozen some of them, she said, and was sending them to a biologist at Ohio State.

This farm wife sounded troubled because in recent years they'd had several calves born without tails, including one that also lacked several inches of the posterior spine. Another deformed calf had just been born, she said, lowering her voice. "This one has an extra leg sticking out of his front leg at the knee."

A woman called from a town north of the Twin Cities and told me her son had seen several deformed frogs. "Is there something wrong here? We want to buy a house."

I sat back, wondering where this would take us. This time it was real. Deformed frogs were not just a freak event at the Ney Pond or in Granite Falls. Suddenly they were showing up across Minnesota, in several other states, even Canada. Why?

Right then, everything seemed up in the air. What among several possible causes of deformities might be acting on the frogs? How could we possibly figure this out? Could different factors be causing abnormalities in frogs at different sites?

On the personal front, I wondered whether Verlyn and I could successfully share a life together given our many layers of personal history. By then, he'd shared with me the tragic story of his first wife's death by suicide and I'd told him about my former husband's massive stroke, which happened when we lived in DC shortly after our second son, Steve, was born; how this eventually curtailed his ability to teach college chemistry it and galvanized me to earn my PhD and work to support our family.

CHAPTER 2

THE FROG CHAMPION

In October of 1995, not long after the discovery of deformed frogs, eighty-four-year-old state representative Willard Munger organized a hearing about deformed frogs and wetlands issues. He invited the school students, their teacher, and me to speak to a mixed assemblage of legislators and citizens. I was honored to be asked and excited to see both wetlands and frogs on the day's agenda. Munger understood the connection, if others did not. He launched a bill in the state legislature to fund an investigation into the frog deformities. At the same time, he promoted new legislation to protect wetlands.

A rumpled-looking, stocky man with a mumbling voice, Munger was a crafty legislator whose entire life's work was aimed at protecting Minnesota's environments. In a country made cynical about government, he remained optimistic that politicians could affect society in positive ways. Over the course of his lengthy legislative tenure, Munger authored and shepherded many laws and regulations related to wastewater treatment, hazardous waste, pesticides, parks and wilderness areas, recycling, energy conservation, air quality and more. Fish recovered in the river that flowed through Munger's home city of Duluth, thanks to the new wastewater treatment plant he'd promoted back in the 1960s.

Munger reminded me of my dad, who was a chemist, not a politician. Munger grew up in humble circumstances, as did my father, and both worked hard all their lives. Dad became a registered pharmacist while in college in California, having assisted in a drugstore in high school. As a druggist, he had put himself through the University of Southern California while he helped support his parents. My father had encouraged me through much of my life, as Willard Munger was doing now in his own way. They were also similar in their occasional curmudgeonly behavior.

In the late 1960s, Representative Munger fought to ban DDT in Minnesota after Rachel Carson had raised awareness about the dangers of pesticides, DDT included, in her pivotal book, *Silent Spring* (1962). In Minnesota, newspapers carried stories of robins quivering and dying from eating DDT-laced earthworms that had consumed the fallen leaves from sprayed trees. At that time, the city was aggressively fogging its abundant elm trees with DDT to kill the beetle that carried the deadly Dutch elm disease. Unconfirmed legend has it that Willard Munger took a basket of dead robins to a legislative hearing to make his case for a ban of DDT. His efforts to ban the insecticide were thwarted because the state's Department of Agriculture passed laws that gave them all regulatory authority over pesticides, rather than the nascent Minnesota Pollution Control Agency, formed in 1967 with the mission to protect the air, land, and waters of the state.

At the federal level, however, the new US EPA, begun in 1970 under President Nixon, assumed responsibility for regulating pesticides, not the federal Department of Agriculture. In short order, the EPA banned the use of DDT nationally in 1972, and that same year the new federal Environmental Pesticide Control Act authorized the EPA to ban the use of pesticides that had unreasonable adverse effects on the environment. By the time DDT was banned, populations of eagles and other raptors had plummeted and many thousands of animals and other birds had died.

In 1962, Carson's *Silent Spring* had awakened the public. People could see DDT's effects in their neighborhoods. They suffered from airborne smog. They knew their rivers were heavily polluted, fishless, even flammable. Expectations grew that the government would step in and take action to control toxic pollution, regulate chemicals, and protect the health of humans and wildlife. The first Earth Day, held in April of 1970, signaled a new era: Congress enacted several major environmental laws, including the Clean Air Act, the National Environmental Policy Act, and the Endangered Species Act.

The Clean Water Act (CWA), passed by Congress in 1972, has a stated goal to "restore and maintain the chemical, physical, and biological integrity of the Nation's waters" and to make the nation's rivers fishable and clean for swimming by 1983 (US EPA 1972; Gross and Dodge 2005; Andrews 2006). To accomplish these lofty goals, pollutants could not be discharged into waters without regulatory oversight by the EPA.

The EPA's Office of Water oversees water quality issues related to surface waters, such as streams and lakes, wetlands, coastal areas, and drinking water. The Office of Water also helps state agencies (such as the MPCA) develop water quality standards largely for chemicals but also for the amount of solids allowed in wastewater discharges. The standards dictate how much of particular pollutants can be allowed into surface waters without harming aquatic life. For example, many fish in freshwater need a minimum level of dissolved oxygen of five milligrams oxygen per liter of water (5 mg/L). If oxygen consistently drops below this, there's a violation of that water quality standard.

Today's EPA has multiple goals, including clean air, reduced greenhouse gas emissions, ensuring safe food and drinking water, and restoring and maintaining various aquatic ecosystems both for human health and the fish, plants, and wildlife that use them. EPA governs the superfund site cleanup program aimed at removing or reducing hazardous waste; it manages the inventory of toxic chemicals, databases that make information available to the public about where chemicals are released into the environment (except those classed as "confidential business information").

In late October of 1995, four of the students from the Minnesota New Country School spoke at Representative Munger's legislative hearing held in southern Minnesota. Sitting at a long table set with microphones, they each told an attentive audience of legislators and citizens what it was like to find deformed frogs at the Ney Pond that August.

"We saw these frogs jumping around and I caught one," said Jeff Fish, age thirteen. "I looked at it closely and I said, 'Oh my gosh, I broke its leg!' And then we picked up another one and it had a broken leg . . . we started to chart down on paper what we were seeing." They were catching them left and right, said Jeff, and it "was weird."

Ryan Fisher, a poised fifteen-year-old, explained to the assembled crowd that they were creating a website and contacting experts on the Internet. They sought information about the chemicals used at the Ney farm. They wanted to find out what could possibly be causing the scary deformities they were seeing, Ryan said, and he made it clear they were doing more than simply collecting frogs.

I summarized our experiences so far and said we were exploring various hypotheses held by scientists for possible causes of the frog deformities—ultraviolet light and parasites as well as chemicals. I explained briefly how ultraviolet rays from the sun might damage the genes that direct normal development in frogs. Leopard frogs lay their eggs in masses that float near the water's surface, I said, and this could expose the newly dividing eggs to ultraviolet radiation.

We also planned to explore evidence that trematode parasites (commonly called flukes) in certain immature stages might embed themselves in young tadpoles and interfere with normal limb development, I said. In addition, we knew that particular chemicals, including some pesticides, could cause developmental abnormalities. So chemical causes would also be investigated. At this stage, no one knew the answer, I said. No one knew if there might be one or several causes of frog deformities. Scientists could only speculate and suggest future research to help explain why this was happening so widely now, in the mid-1990s, and not before. We were entering new territory. I acknowledged our shared concern: that whatever was harming the frogs might also harm humans.

I made a plea for better regulatory protection of vulnerable, small wetlands that frogs use to reproduce. Small wetlands are too easily removed from the landscape, I said, yet they are prime frog habitat.

After a couple of presentations about wetlands issues, the hearing opened up for comments from citizens. There was no shortage of opinions. One man stood up and expressed concern about government regulation of wetlands, saying, "This is America, you don't go in and take land without compensation." He felt strongly that individuals have the right to do anything on their own property, to wetlands or otherwise. These private rights, he said, should not be trumped by environmental laws imposed by the government.

A citizen stood up and called wetlands "the placenta of our existence. Wetlands have an eternal majesty. They are the rainforest of diversity in Minnesota. If we lose them, we lose part of ourselves." Another proclaimed, "if we're going to survive the next two hundred years, it will be because we gave priority to the frogs."

The next day, Cindy Reinitz e-mailed me: "What a great experience!" After the hearing, Representative Munger had asked her if she knew what she and her students had done there that day.

"You just passed a wetlands bill," he'd told her.

People understand pretty well what streams and lakes are, but wetlands are more complex, mysterious even. Some would ask us what exactly is a wetland, thinking mostly of cattail-bordered ponds with open water and a few ducks floating about. There are several broad categories of wetlands: marshes, bogs, wooded swamps, short-lived temporary ponds (that are often dry in fall), vegetated fens, and wet meadows (that can look grassy). In some, the water isn't obvious, at least from a distance.

The US Fish and Wildlife Service (USFWS) defines wetlands as areas where regular saturation by water is a key element in determining the types of plants and animals that can survive in wet conditions (Cowardin et al. 1979). Wetlands also have specific kinds of underlying mucky or mineral soils that result from the natural lack of oxygen and the deposition of organic matter derived from partly decomposed plant material. Both the kinds of soils and the types of wetland-dependent or -tolerant plants are used as indicators of the presence of a wetland and its boundaries, even if surface water is lacking (US Army Corps of Engineers 1987).

Some wetlands dry down periodically, others, with permanent water, look more like lakes. The USFWS defines wetlands as having a maximum depth of 6.6 feet (or 2 meters) under certain conditions. This is the general limit of water depth that most emergent wetland plants, such as cattails and bullrushes, can tolerate.

To develop valid scientific measures of wetlands condition, Mark and I chose to work on marshes and shallow water ponds, or, in our lingo, "depressional" wetlands. Such wetlands range from seasonally flooded to permanently ponded sites that serve as important habitat for reproducing waterfowl and frogs. If a wetland had polluted mud and water or had been physically disturbed, then we'd expect to find fewer species of sensitive, wetland-dependent plants and invertebrates. If it were degraded, we'd record negative changes in the biological communities.

By 1970 drainage projects had removed more than half of the nation's wetlands, thus diminishing the habitats that support wetland-dependent wildlife (including frogs) and leading to increased surges of water flowing into ditches, tributaries, across floodplains, and into rivers. In some states, for instance, all of Iowa and the southwestern region of Minnesota, drainage has eliminated more than 90 percent of historic wetlands (MPCA 2006). Conservationists began to call upon the federal government to prevent further wetlands losses and begin a new effort to restore previously drained wetlands.

In the late 1980s a broad-based national task force made numerous recommendations to slow the continuing loss of wetlands (Vileisis 1997; Pittman and Waite 2009). The committee espoused a new policy that called for a "no net loss" of wetland functions, such as filtering pollutants, modulating floods, and recharging groundwater. In 1988, George H. W. Bush promoted the concept of a no-net loss policy for wetlands during his presidential campaign.

This new approach to wetlands protection would eventually require the replacement (or outright creation) of wetlands to compensate for natural wetlands that would be destroyed for various reasons. But scientists have raised concerns that so-called replaced or restored wetlands often do not support the biological diversity of natural wetlands (National Research Council 2001).

In one view, replacing at least some of the *physical* functions, such as flood prevention or settling out pollutants, of wetlands counted as much if not more than replacing lost biological communities. There was an attitude that said, "Build it and they will come." Create the pond and plants and animals will colonize it in the future. But created wetlands may lack the rich array of plants and animals or the organic mud that is infused with seeds and seasonally dormant eggs of invertebrates, mud long deposited from decaying wetlands vegetation. And rarely does anyone check whether the natural life has been restored (although see Dodson and Richard 2001).

More recently, the second President Bush proclaimed that the no net loss of wetlands had been achieved over the period 1998–2004 based on a government report (Dahl 2006). But the gains in acres used in the net loss calculations were contrived: man-made golf course ponds, recreational and decorative ponds, and even storm water retention basins and sewage ponds were counted as wetlands. Such "wetlands" have some physical but few biological functions (National Research Council 2001).

In 1991 Minnesota passed its new Wetlands Conservation Act, a complex law fully enacted in 1994. Its goal was to achieve a no-net loss of the "quantity, quality, and biological diversity of Minnesota's existing wetlands" (Minnesota Administrative Rules, Chapter 8420.0100). To me, aiming to protect the biological diversity of wetlands was important and hopeful.

The goal was lofty, but in reality the state's Wetlands Act allowed dozens of exemptions. Exempt from regulatory oversight were small, temporary wetlands on agricultural lands; up to two acres of sedge meadows and shrub wetlands on farmland; and wetlands affected by various public improvement projects like transportation, pipelines, and power lines. Monitoring whether biological diversity has been sustained would rarely if ever be done. The new wetlands law would allow what's called "out of kind" replacements, which meant that one could destroy an ancient bog and replace it with a flooded cattail marsh.

When hearings on Minnesota's proposed Wetlands Conservation Act were being held in the early 1990s, I decided to testify in favor of protecting small, seasonal wetlands because they support unique biological communities, not just certain frogs that needed them for reproduction in early spring, but also a variety of invertebrates as well. Fairy shrimp, for instance, are ancient crustaceans that exist only in seasonally flooded, temporary pools because their resistant eggs require a dry-down period. As I spoke about the biodiversity supported by seasonal wetlands, I approached the presiding administrative law judge, Alan Klein, and handed him a glass vial of preserved fairy shrimp. He rolled the vial in his fingers and gave me a very quizzical look. We both knew the chances of the new law protecting the smallest wetlands were slim to nonexistent.

Representative Munger later publicly praised the students for their presentations at his hearing that fall and said, "The frog is a warning to this civilization as the canary was to the miner." The students who spoke up, he said, look to the government to help protect the environment for their future. Early in his life, Munger's grandfather had urged him to go into politics so he could help save the woods and wetlands of Minnesota where he grew up. He had a consistent goal to work to protect the environment during all of his life.

In contrast, while I loved the streams and ocean of my childhood in New England, my belief in the responsibility of government to work for the public interest and my desire to protect the environment didn't develop until much later in life. I was pretty insulated in college during the 1950s, only vaguely aware of the burgeoning civil rights movement and the efforts of college students who risked their safety—and even their lives—for a larger goal. And in college I was enamored of the new molecular biology, not ecology nor the effects of human activities on fragile species.

Not until my first husband and I moved to Washington, DC, in the 1960s did any activist urges awaken in me. By then I fully understood the need for new laws to protect people's civil rights. We marched against the Vietnam War. I organized a small demonstration of women and their children to stand on street corners with STOP THE WAR signs, my toddler son, Erik, holding up his PEACE sign from his stroller. In the late 1960s, my husband and I were rocked by the assassinations of Martin Luther King and Robert F. Kennedy, by the riots and conflagration in downtown DC after MLK's tragic death.

We were in Washington, DC, during April of 1970. I'm embarrassed that I can hardly recall the major demonstrations held in the city for the first Earth Day because our family was in crisis mode because of my husband's disabling intestinal illness and major stroke. In May of 1970, Lon had recovered enough to return to the classroom. He was tear-gassed out of his class that month when a large demonstration erupted in the streets close by, not for Earth Day, but an outpouring of rage over the deaths of students shot during an antiwar protest at Kent State University. I remember the helicopters flying overhead, the armed soldiers standing at intersections. DC was a city seemingly at war.

From DC we moved to Luther College in Decorah, Iowa, a town nestled in the unglaciated hills of the northeast corner of the state. The boys—barely three and five—were free to roam, to collect ancient fossils in a nearby dry streambed, find salamanders, and build forts at the edge of the woods. Decorah was a reprieve for us from the personal and public stresses of the Washington, DC, of the late 1960s. We had three years of relative peace in Iowa before we had to leave.

We moved from Iowa to Seguin, Texas, where my husband had landed a one-year teaching stint at Texas Lutheran University. By then I knew that I would have to retrain, find work, and support our family. My experience sampling polluted and pristine rivers during an aquatic ecology class I took at Southwest Texas State University fostered my commitment to study the biological effects of pollution.

That change of heart, that epiphany by Texas rivers, had happened in the mid-1970s, when the newly formed EPA was taking heroic steps to regulate pollution. That government could work to protect the health of living systems influenced my future direction.

After our year in Texas our family moved north to Minnesota in the mid-1970s, where I did graduate work for the PhD in zoology at the University of Minnesota, with emphasis on aquatic ecology and freshwater zooplankton such as *Daphnia.* I worked a responsible job at the university while taking courses and conducting experiments on *Daphnia* in the lab. When I started my program, my sons were in elementary school; when I finished they were in high school and stood taller than I did. By then my husband and I had agreed to a separation that eventually led to our divorce in 1983, after twenty years of marriage.

In the 1980s I had landed a temporary teaching position in the Biology Department at St. Olaf College that lasted for five years and put a lot of miles on my car commuting to the college. After St. Olaf I took a postdoctoral fellowship at the University of Minnesota, doing research on the molecular biology of mosquitoes. Having a more solid background in the newer biology would deepen my teaching of undergraduates, I felt, and improve my chances for another college teaching position. Cellular and molecular research were producing new, exciting results at that time and helping to explain or illuminate a range of issues in biology.

I was drawn to this new approach to basic research, yet I wanted to stay with my goal of working with freshwater invertebrates, to understand how pollution harmed them, to work on whole organisms in the natural environment. Way back, not long after college, I'd assisted in a biochemistry lab at Tuft's University Medical School where we isolated subcellular particles (ribosomes) from rat livers and researched how ribosomes help synthesize proteins. Taking cells apart and learning how they work at the most basic level was fascinating work.

When a job arose at the Minnesota Pollution Control Agency in 1989 in the middle of my postdoctoral work, I grabbed it, even though it was a short-term grant and I'd have to leave my postdoctoral research behind. I began working at the MPCA on a research project aimed at understanding the ecology of organisms that inhabited sewage ponds. The goal of my part of this project, which worked in cooperation with University researchers, was to learn why some of the ponds had healthy *Daphnia* populations, tiny crustaceans capable of consuming blooms of algae promoted by the nutrients in sewage, while other ponds did not. Those that lacked *Daphnia* released green, algae-ridden water in their outflows that discharged into rivers, sometimes triggering water quality violations (Helgen 1992).

My doctoral work had been on *Daphnia.* But I had no idea that my graduate research might lead me, wearing long rubber gloves and boots, to row a small boat on wastewater ponds that were laced with globs of raw sewage so I could net and study the zooplankton. Amazingly, neither I nor my student assistants got sick while handling raw sewage water.

I brought with me to the MPCA a small grant, one I'd been previously awarded, to assemble information on the state's freshwater invertebrates. This grant became the springboard that would lead to my work on monitoring the health of wetlands by analyzing their invertebrate life. By then, documenting pollution with biology was what I really wanted to do. The EPA began awarding me small grants for three months, then six, to cover my salary so I could begin to develop biological indicators for wetlands. Although temporary, I hung on at the MPCA. When Mark Gernes, with his expertise in aquatic vegetation, joined me in that effort, we worked together to develop the tools to quantify pollution in wetlands by measuring the changes in their biology.

The day after Representative Munger's legislative hearing in the fall of 1995, I was back at work trying to get caught up on our long-delayed wetlands work: data analysis, reports, and a book chapter, all with due dates. We had barely begun to develop scientific methods for sampling the organisms and plants in wetlands and analyzing what we found. We had to stay on top of it and complete the regular progress reports required by the EPA or we'd lose our funding.

The EPA wanted researchers to develop scientifically valid, biological measures to test levels of pollution or degrees of degradation to wetlands. In the future, with these tools, states could monitor and assess the quality of their wetlands. The ultimate goal: to use well-developed, biological rating systems to assess wetlands water in the same way that the analysis of many chemicals (such as ammonia) and other factors (turbidity, dissolved oxygen) are used to rate the water quality of streams. The new research fostered by the EPA would eventually lead to the development of "biocriteria," biologically based rules that could be used for assessing and regulating pollution (US EPA *Wetland Bioassessment Publications*).

Mark and I researched the communities of aquatic invertebrates and water-dependent plants. In other states, researchers, also supported by the EPA, would explore using species of algae, birds, fish, and amphibians as possible ways to measure wetlands health. Within each category, multiple species were analyzed, both those deemed to be sensitive to pollution and those that tolerated it. The broader the range of species within a category the better. Using a single indicator species—such as one kind of plant or one species of dragonfly or frog—would not be robust enough to detect a range of pollution in wetlands.

It was premature to judge whether the presence of deformed frogs indicated pollution in wetlands. Developing the multispecies biological indexes was our long-term goal for evaluating wetlands. Finding the cause of deformities in frogs was also important. But could we sustain added work on frogs without additional help?

In fall of 1995, after the hearing sponsored by Representative Munger, I had my annual job review. My supervisor praised me for hard work but then told me our section manager didn't like the way I worked. "Impulsive" was the word he used. Apparently, my involvement with the frogs was rattling management's nerves. Mark and I believed our work was important to the agency, as did Representative Munger. I wished our bosses felt that way too. Was it impulsive to chase after malformed frogs?

The veiled reprimand didn't quell my drive to uncover what caused the frogs to develop so weirdly. How could we prevent such deformities from happening again in the future if we didn't know what was responsible for what we'd just seen? Was it natural or caused by humans? Even if we didn't succeed, I felt we had to try to figure it out, whatever that would take.

During the fall of 1995, after the students' discovery and the flurry of media attention that followed, scientists had been generously donating time and analysis at their labs. In Montana, Johnnie Moore, a geology professor of my son Steve's, was analyzing pond samples for toxic selenium; Bob McKinnell at the University of Minnesota was dissecting frogs to examine their internal structures; a University of Minnesota chemist was testing water samples taken from shallow groundwater beneath the artificial islands in the Ney Pond; the state Department of Agriculture's chemistry lab analyzed a few samples for pesticides. All this at no cost to us.

We'd had lots of cooperation that first year, but soon we'd need to pay for various kinds of analytical work And that meant pushing the MPCA to find new funding.

I began to ask MPCA managers if the agency could seek state funding to allow us to investigate the frog deformities in the coming year, 1996. Mostly I ran into a brick wall. "To get the governor to look at an issue that was not presented to him prior to this [the finding of deformed frogs in summer] is *not* going to happen," said the Water Quality Division director. Unfortunately for us, the deformed frogs had emerged too late for the agency's budget planning process.

I became acutely aware that state agencies, as with federal agencies, fall under the executive branch of government. The MPCA worked for the governor, Arne Carlson at the time, who had the right to review and adjust the agency's budget before he sent it on to the legislature for approval.

One day, I received an e-mail from our Water Quality Division director warning me about the Munger bill that proposed funding for the MPCA to work on the deformed frogs. I needed to be aware of the implications of *going around the process* on this, she wrote. Mark and I were *not* to lobby any legislators for money for the frogs, not even Representative Munger. I was told to inform Munger, who was by then actively and independently seeking state money to investigate the frog deformities, that the MPCA *could not* support his proposed bill. The agency can't afford to "rob Peter to pay Paul for the frog study," our director cautioned. I slumped in my chair, discouraged by this attitude, by this apparent lack of desire to tackle what was wrong with the frogs. Why was that? I wondered, gazing out the window toward downtown St. Paul. It seemed crazy to me that our agency would stubbornly refuse any money brought to them by a state legislator.

The suggestion that getting money to investigate deformed frogs might jeopardize other staff's funding disheartened me. The MPCA's managers didn't seem to dare (or want) to suggest that this newly emerging environmental problem might merit a separate request for additional funding from the state. Were they under pressure from the agency's top administrator, the commissioner?

The state's budget runs on a two year, or biennial cycle, that starts July 1 every other year. First the MPCA designs its budget, making decisions on what it needs for the wide range of work done by staff, everything from monitoring to education to enforcement of regulations. Then the budget proposal is sent to the governor for approval before the legislature amends the budget and passes the final funding bill, sometimes not until late May. The budget is, to a considerable degree, politically controlled. The timing—a new budget cycle starting in July, makes life complicated for staff doing seasonal work. One can't plan without knowing in advance what funding will be available, for instance, for hiring seasonal workers or ordering supplies. Plus the system for forming contracts that might need to be in place by early July requires a good two to three months, sometimes more.

Having the funding for our wetlands work come entirely from the EPA was an advantage for Mark and me. The federal fiscal year begins in October, allowing us time to plan our summer fieldwork well in advance, through the winter. Federal funds to state agencies are usually targeted for specific projects, also an advantage for Mark and me, because otherwise our funding could be vulnerable to the "reallocation" process, in which managers can shift money from one account to another. We'd later discover our grants were not immune to such tactics.

I was beginning to see a jarring disjunction between the MPCA's stated goals of environmental protection and the realities of its actions. Huge feedlots had little oversight, a refinery's contaminated groundwater was allowed to travel toward the Mississippi River. More emphasis was placed on what was called "voluntary compliance" by industries to the state's water quality rules than on outright regulation. Historically, this softer approach hadn't worked, because most companies have little motive to police themselves when it comes to pollution. Perhaps it worked in some cases—encouraging a polluter to do something that would avoid incurring a fine.

During that fall, the MPCA began promoting a new strategic direction, one that appeared more economically than environmentally oriented. Managers raised questions about the "cost effectiveness" of regulations. How much would reducing some specific kind of pollution cost compared with the environmental results, the benefits? One staff person pointed out that the costs to the environment resulting from human activities weren't even being considered in the cost-effectiveness equation. Managers talked of providing good "customer confidence" and "customer service" among the industries and other entities that interacted with the agency.

In a meeting I asked, "Aren't our customers broader than the regulated communities? Don't our customers include lakes and forests, air and soils, fish, birds, and even frogs?" A manager gave me a quizzical look. Later I was told that the concept of customer service meant doing a better job of explaining MPCA regulations and why they needed to be enforced. That sounded reasonable. But to some staff it appeared that our agency was slipping away from its mission of protecting the physical and biological environment that fostered both human and ecological health.

In November of 1995 Bob McKinnell and I were invited to talk at the University of Minnesota about wetlands and frogs. As we set up our slides for the presentations, I was surprised and excited to see Representative Munger walk into the lecture room with his aide and a reporter from the Minneapolis *Star Tribune*. For my talk, I reviewed our findings of deformed frogs in Granite Falls, at the Ney Pond, and at the residential property north of Litchfield. I repeated my plea to protect small wetlands that frogs need for reproduction. I elevated the value of using biological indicators to measure environmental health and the need to investigate the frogs.

During the discussion after our talks, Willard Munger stood up beside Bob McKinnell. Together they delighted the audience by swapping stories about "frog picking." As a boy Munger had earned money by collecting frogs. All he had to do was drag a gunny sack along the edge of a pond and easily fill it with frogs, he said. There used to be thousands of frogs to sell, the small ones for fisherman and bait dealers, the large ones for commercial froggers who wanted the legs for restaurants. Munger's views on frogs had obviously changed since he was a boy out collecting them to earn some money.

"Nowadays, you can't fill a bucket with frogs," the elderly legislator said, looking angry. "Frogs are the number-one indicator of a healthy environment, and we had better change our direction or we are in real trouble."

Amen, I thought.

Later I would learn that indeed the ordinary leopard frog was in trouble across the United States and Canada. Representative Munger's stories—and those I heard from people who lived in rural areas—foreshadowed a real decline in populations of leopard frogs (see epilogue).

Representative Munger succeeded in obtaining funding for us to work on the deformed frogs and money to educate and involve school children. He raised the issue to the media and focused public pressure on getting the government to take action. He fought against MPCA leaders who felt it was inappropriate for the agency to tackle research on frogs. Munger had served more than forty years in the state legislature, up until he died in 1999 at the age of eighty-eight.

People called him Mr. Environment. To me, Willard Munger was the Frog Champion.

CHAPTER 3

LEARNING CURVES

One thing I knew: to find clues to what caused the deformities, we had to focus on the frogs themselves and the places where they spent their time throughout the year. The most likely scene of the crime lay within the ponds, where developing eggs would be exposed to pollutants or other agents in the water. But what about places where the adult frogs, females especially, went to feed or spend the winter? Could pollutants from those habitats get into their bodies, penetrate the eggs, and cause abnormalities?

I was sitting at my desk, contemplating how to deduce where the Ney Pond frogs overwintered, when a staff hydrologist strode into our work area. Normally a low-key kind of guy, he had a serious look on his face.

"Hey, Mark and Judy, looks like the lab at the U [University of Minnesota] has picked up some radioactive iodine at the Ney Pond," he said.

"You're kidding," I said. "Where from?"

That summer, he'd extracted water from a few feet below the soil surface of one of the pond's tiny created islands—"pore water," he called it. The lab was also checking into possible evidence of uranium in the sample, he said, but they doubted that result was real.

How could radioactive material have ended up at the Ney Pond? I immediately thought back to the fallout that spewed all over the United States during the years of above-ground nuclear bomb tests at the testing grounds in the Nevada desert during the 1950s and 1960s. Iodine is incorporated into thyroid hormone, the hormone that drives the transformation from tadpole to four-legged frog. Could radioactive iodine deposited in the thyroid glands of tadpoles alter their limb development? This was another possible cause that no one had yet considered.

"If radioactive fallout from atmospheric bomb tests landed around the pond in the sixties," I asked, "wouldn't the radiation have decayed by now?"

Later I looked it up: strontium 90, a radioactive isotope from nuclear fallout, has a half-life of twenty-nine years. Similar to calcium, it lodges in calcified tissues, muscles, and bone marrow, where it may cause bone cancers and leukemia, according to the Centers for Disease Control. I remembered news reports of children whose shed baby teeth were found to contain radioactive strontium, ingested from the milk of cows that grazed on fallout-contaminated grass in the early 1960s. Iodine 131, also in fallout and known to accumulate in the thyroid gland and cause thyroid cancer, has a half life of only 140 days. It too was found in cow's milk in the 1950s and 1960s. If radioactive iodine got into the soil of the island in the Ney Pond it would have had to be quite recent, not from the bomb tests more than thirty-five years earlier.

"What about the meltdown of the nuclear power plant in Russia?" Mark countered. "That was in the late 1980s." We remembered that disastrous event: the accident at Chernobyl in 1986 exposed thousands of Russians and others in the region to high levels of radioactivity. Crops and land were contaminated, probably for decades to come. One study estimated that fifty thousand new thyroid cancers, especially in children, may have been triggered by Chernobyl. However, the iodine would have certainly decayed by 1995.

But radioactive material in Minnesota? At the Ney Pond? It seemed very unlikely.

It was a long shot, but I had to pursue this, yet another tangled path in the deepening forest of possible causes. I called the Minnesota Department of Health and asked a staff person for their data on radioactive fallout in the state going back to the 1950s.

"We don't monitor fallout," he told me.

"Not anywhere?" I asked, incredulous. What about Minnesota's two nuclear power plants located south and north of the Twin Cities?

"No, I'm sorry."

How could we narrow the field of causes when more possibilities kept popping up? It alarmed me how little scientists seemed to know about which of many possible factors might harm frogs: thousands of untested chemicals, a few biological agents, ultraviolet light, and now—radioactive iodine? There were too many possible culprits to chase down and too little information about them. It would be so much easier if we had a single suspect to pursue—like the toxic selenium at the Kesterson marsh.

What were the frogs at the Ney Pond most likely to encounter that could harm development? The pond was surrounded by agricultural fields and fed by four tile lines draining water from cropland. Analysis of farm chemicals was of prime importance, even if radioactive iodine was in our thoughts for the moment.

Could the frogs pick up toxic chemicals from the insects they ate in the vegetated areas around the pond or by absorption through their permeable ventral skin from the soils and landscapes they migrated across? What about the water where they spent the long winter? Could they absorb pollutants there?

By late October 1995, most of the frogs had left the Ney Pond, presumably heading for deeper water by trekking down the bluff to the Minnesota River. I was pretty sure the pond itself was too shallow. If it didn't freeze all the way to the bottom, it was likely the water would become depleted of oxygen during the winter. As microbes decompose the algae and plant material in the water, they consume dissolved oxygen that frogs need to survive the winter. Leopard frogs have to enter the deeper water of lakes or rivers, which retain more oxygen than shallow ponds. They don't burrow into mud, as I had earlier thought; instead they sit on the bottom, sluggishly awake all winter in near freezing water. Other species of frogs, such as tree frogs and wood frogs, which make their own antifreeze and can survive in a torpid state under debris on land, would not be exposed to pollutants in the muds or waters of rivers and lakes.

"How far might a leopard frog migrate to its wintering site?" I asked John Moriarty, a coauthor of the 1994 book *Amphibians and Reptiles Native to Minnesota* (Oldfield and Moriarty 1994) and a great resource on all things related to reptiles and amphibians and other local wildlife.

"Oh, maybe half a mile," said John offhandedly. That seemed a long ways to go for such a small animal, I responded. Much later a scientist would demonstrate that the leopard frog could travel up to a mile overland (Rorabaugh 2005).

"They have to get to the deeper water for the winter," John said, his enthusiasm over frogs building. "Usually they go to a lake or a river." Fish die in shallow wetlands, a good thing for frogs, either from icing or oxygen deprivation. Frogs too will suffocate, John said, if they can't get enough oxygen from the water to "breathe" through their skin all winter. We both knew that too many shallow wetlands had already been drained to create cropland or filled for urban developments. Sometimes called *isolated wetlands,* such small ponds are among the most vulnerable to loss and the least protected by state and federal law. Yet leopard frogs depend on them for reproduction.

We talked briefly about climate change, a topic emerging in discussions among scientists but not yet in the public eye. We agreed the early spring breeding frogs, like wood frogs, that use the smallest and most temporary shallow wetlands would be in the greatest danger from global warming. The seasonal wetlands will be the first to dry up. Frogs can migrate, but there's a limit to how far they might go to find suitable breeding habitat. Their fatal flaw, if you can call it that, is their need to lay their eggs in shallow ponds, preferably those lacking fish.

To find deep water, the frogs would have to travel a considerable distance from the Ney Pond down to the Minnesota River. Was this too far? I copied topographic maps of the areas around the two wetlands where we'd seen deformed frogs: the Ney Pond, on the east side of the river, and another wetland across the river, not far from Henderson. I taped the maps together, then penciled one-half mile circles around each pond.

"Mark, look at this," I said, showing him my map. The two half-mile circles intersected right in the middle of the Minnesota River. "The frogs *could* migrate that far," I said.

I mailed a copy of the map to Cindy Reinitz, who was driving around the area near the river that fall, looking for evidence of migrating frogs crossing the roads. Could Cindy sleuth out where they overwintered?

In October, Cindy led her students to a dirt bluff by the river's edge. One of the boys started kicking at the eroded bank, and suddenly frogs leaped out! Excited, students peered into cracks in the dirt and saw more frogs hunkered several inches down in the crevices. I was happy to hear of the students' discovery: an important clue that frogs headed to the river.

Bob McKinnell had told me how frogs marshal at the shore of their overwintering sites. "The water has to be cold enough before they go in," he said. "You know how I test 'em to see if they're ready to take the plunge? I throw a frog out into the water. If it does a one-eighty turn and swims right back to shore, then the water's not cold enough." I could tell by the glint in his eye that Bob loved telling this story. Unlike humans, frogs preferred their winter water cold.

Perhaps the frogs were waiting until the water cools enough to slow down predatory fish, I speculated.

I looked into Walt Breckenridge's 1944 book, *Reptiles and Amphibians of*

Minnesota. He described frogs lurking in the open flow of a stream below a small dam during winter. The sandy bottom "was literally paved with hundreds of closely crowded leopard frogs," he wrote. "They were not entirely dormant, for every few seconds, one would float up and swim slowly to a new situation."

Art Straub, a naturalist and teacher from Henderson, was sure that frogs wintered in the river. He'd seen local men out on the ice, spear-fishing in the open river flowages in winter, a sport Art called "promiscuous fishing," in which anything that moves was speared and dumped dead onto the ice, frogs included. Afterwards, eagles would swoop in for the carcasses that lay strewn together with the trash and beer cans left behind.

That fall, Cindy Reinitz and her students gave a talk about the deformed frogs at St. Olaf College and met biology professor Eric Cole, an enthusiastic scuba diver. Experienced in winter dives, Cole had seen frogs underneath ice, but not on the bottom, he told me. Instead, they hovered just below the ice layer, as if trying to get out. He offered to do an ice dive at the Ney Pond in December to check if any frogs wintered there. When he did, he found only one.

Late in the fall of 1995, Mark and I had been scouting the riverine pools in the river valley for the likeliest place that the Ney Pond frogs might spend the winter. A long, wooded pool with river water flowing in one end and out the other looked like the closest site below the bluff. The St. Olaf ice divers, Eric Cole and a computer guy, John Campion, agreed to attempt a winter dive there. By then I privately called them "the crazy St. Olaf ice divers." The idea of going into frigid flowing water (or a lake) through a hole in the ice and risk getting lost or disoriented, seemed insane. Clearly, I was not the risk taker I wanted to think I was.

On a relatively balmy day in late December (in the low twenties with no wind), teachers, students, and a local news reporter met Eric Cole and colleagues near the river's edge. We scrambled down a short bluff to the shore of the large ice- and snow-covered flowage pool. The divers, garbed in black wet suits, hoods, and goggles and looking like creatures from outer space, lugged air tanks and a portable chain saw out to the center of the pool.

Cindy and I, bundled up in bulky down jackets and knit wool hats, stood by in our felt-lined boots. We watched as Eric started drilling a large hole in the ice with the noisy chain saw. Out came heavy, triangular chunks of six-inch-thick ice. Once the access hole was ready, Eric slipped slowly down into the water with his air tanks and disappeared. In a matter of minutes, his rubber-covered head popped back up through the hole. He was taking off his tanks.

"Eric, I hope you're not going under without air," I protested.

"No," he said, "The water's too shallow, there isn't enough room under the ice. I have to carry my tanks sideways." I felt a bit edgy about this venture. Bob McKinnell had said he wanted no part of these plans, because he knew of a diver who tragically lost his life diving under ice. This was dangerous.

While standing by, Cindy and I discussed the ups and downs of our lives since getting swept into the deformed frog problem. She had issues with a disgruntled parent who'd wrongly accused her of not protecting a student who was out wading in the Ney Pond on her own with her family on a weekend. Media requests were putting pressure on Cindy's time at school and after hours as well. I struggled with my reluctant managers, who had made it clear that deformed frogs were not part of the agency's plans and mostly caused them headaches. I shared my concerns that my new relationship with Verlyn might suffer because so much of my time and energy went into my work.

Stamping our feet, we gazed across the frozen pond and watched Art Straub and two students scouting along the edges. Suddenly from a distance they shouted and hooted. Then a triumphant yell, "We found frogs!!"

We joined them at the edge of a dark-looking, open seepage area bordering the shore of the pool. I stepped cautiously on the ice toward the open water and looked in. Sure enough, there were frogs, quietly splayed out on the bottom and moving in slow motion. I smiled. Once again, the students, and a teacher like Art who seemed to know everything about the river, had led us to the frogs. At the end of the pool, we spotted more frogs lying in the open water that flowed back to the main river channel. By then, the St. Olaf divers had emerged from under the claustrophobic ice. It had been too murky under there, Eric said. They'd tried to feel around for frogs but hadn't encountered any.

We netted out and examined frogs from the open water areas at the riverine pond. None of them appeared deformed. Did this mean that the deformed frogs were so handicapped that they couldn't make it down the bluff to the river? Had they frozen to death along the way or been eaten by predators?

In graduate school at the University of Minnesota, one of my zoology professors, Bill Schmid, studied the ability of animals like frogs, snails, and insects to tolerate freezing. Schmid knew that certain species of frogs, like the wood frog, spend the winter under leafy debris in the woods, not in the safety of deeper water. How do they survive Minnesota's subzero winters? He discovered they create their own antifreeze chemicals, chemicals that allow their tissues to freeze without killing the frog or rupturing the integrity of tissues and cell membranes. He could freeze these frogs in his lab, then, like some kind of mad scientist, bring them back to life by letting them thaw gradually (Schmid 1982). Biologists have speculated that frogs that tolerate freezing might have survived better during historic periods of glaciation in North America. Sometimes being cold-blooded has some advantages, especially if you can make your own antifreeze. Leopard frogs, however, can't remain on the land in winter, at least not in Minnesota. They must seek the protection of relatively warmer water to survive the deep freezes.

Later, Bob McKinnell told me how some commercial froggers cut holes in the ice to attract frogs. "They get more oxygen from the open water, I suppose," he said. Presumably, the froggers were selling the bigger ones ("lunkers," as Bob called them) to biological supply houses or restaurants.

We had confirmed that leopard frogs lie around under the ice for five long winter months in contact with the polluted water and bottom sediments of the Minnesota River. Later the students would demonstrate where the frogs moved down the bluff from the Ney Pond to the river. I thought again about the eggs the female frogs hold in their oviducts all winter until the eggs fully mature in spring. Could toxic chemicals in the water or sediments penetrate their skin and possibly harm the eggs? Or were the eggs safe in this winter refuge, not absorbing any pollutants?

One of the reasons to do the ice dive, other than confirming that frogs spent the winter in the river's flowages, was to collect more female frogs for Bob McKinnell at the university. Earlier that fall, the students had collected a couple of large females they named Josie and Mo from the vegetated area around the Ney Pond. We had taken the pair to McKinnell's lab so he could fertilize and test the eggs to see if they developed normally.

Josie yielded eggs that had fertilized well, but Mo died unexpectedly in the lab. Bob had seemed quite troubled about Mo's flaccid condition. He couldn't understand why she had died. Josie's eggs had high rates of fertilization, with three different sources of sperm, one from an East Coast frog, Bob said. Then he gave me the bad news: "All of Josie's early embryos were so deformed that none of them survived to hatch out from their jelly coverings."

Bob, who had fertilized frog eggs thousands of times for his research over the years, had never seen anything like this. "Even the very earliest blastula stages [the first ball of cells] were deformed," he'd said, his voice rising. Why had none of Josie's embryos hatched out successfully? What was wrong with the eggs? We couldn't say too much with results based on so few frogs. But Bob's report was, nonetheless, quite disturbing. Would eggs from other female frogs that site do better? What would next summer's crop of new frogs show us?

I remembered when we fertilized frog eggs during college and observed their elegant development; how the eggs divided so neatly, at first into smaller cells near the top and larger ones lower down, then to a hollow ball of uniformly tiny cells. In embryology lab, we learned how primordial cells moved and formed layers for future muscle, gut, and nerve tissue; how the lumen of the intestine formed, the nerve cord rolled up and sealed shut, the brain and eyes established in the developing head of the tadpole larva. At that time, developmental biology was transitioning to the new era driven by exciting research in molecular biology. How do these dividing cells, each receiving the same chromosomes and genes, end up becoming specialists like muscle, skin and nerve cells? That problem, the mystery of cell differentiation, led me to major in biology rather than chemistry. And molecular biology stood poised to begin solving it.

Could today's scientists work to understand how this elegant process of development had been so terribly corrupted in these frogs and by what?

Bob McKinnell had seen deformities in the very earliest stages in the eggs of Josie and Mo from the Ney Pond. Would this repeat itself with at least one or two more females, which we collected from the river's ice-covered pool? If those females' eggs had problems, it could point to the water quality of the river or contamination in its sediments. Or it could mean that the eggs carried harmful contaminants the females had previously absorbed from insects they ate during summer or from the landscapes they hopped across.

Later that winter, Bob called me with the results. "The eggs from those females you collected in the river? Their embryos are in even worse shape than Josie's! They were all grossly defective," Bob said. Researchers knew that certain chemicals could disrupt the early cleavage stages of a fertilized egg. An antibiotic, actinomycin D, and a drug, lithium, were known to disrupt and distort the earliest stages in sea urchin and frog embryos. But what distorted the eggs that Bob had attempted to rear in his lab?

Hearing his disturbing results sharpened my commitment to pursue the cause no matter what. We had to investigate this peril. But I had a strong sense of foreboding that my agency had no desire to investigate what was causing the disturbing abnormalities. There had been whispers that the MPCA would not seek future funding to support work on the frogs. My bosses had voiced unease about my investigation. How much could I pressure them to tackle this environmental issue? At what personal cost? My only hope was Willard Munger, who had so quickly embraced the cause of the deformed frogs and encouraged me and the students to stay involved. Representative Munger wanted an investigation. I'd have to do whatever I could.

By 1996 the enormity of my workload and the limits of my knowledge hit home, especially after witnessing frog deformities of every imaginable kind the preceding summer and having no real sense what the cause or causes of this apparent epidemic might be. Pressures built and demands for answers increased. Calls from media people, citizens, and scientists absorbed much of my time during the normal work day. I couldn't properly analyze our wetlands data or write overdue reports about our research for the EPA. I frequently stayed late and went in on weekends but still couldn't catch up. How could I be developing biological indicators for wetlands while simultaneously helping to unravel what was happening to the frogs?

My work consumed me. I'd had a long, uphill climb to earn my reentry PhD, had worked at temporary college teaching jobs, researched the impact of pesticides on ponds with the EPA, then landed my job at the MPCA, where I worked to promote the monitoring of wetlands water quality. Now I had taken on a new role, one that seemed overwhelming at times. Something was terribly wrong—I had to find out what, or at least try

A mind-boggling array of hypotheses for causes of malformations in frogs was landing on the scientific table as we headed into 1996. These included various pesticide products (herbicides, insecticides, fungicides, worm killers); chemicals that could disrupt hormones that play important roles during early development; toxic metals like selenium and mercury known to cause deformities; ultraviolet light; even radiation, parasites, and other biological agents.

What could have caused this peril in our ponds? Clearly no one scientist could tackle this question alone. The frog problem cried out for a full sharing of ideas, a fruitful cross-fertilization among many experts. We needed many researchers of various backgrounds to work together to seek answers and pin down the perpetrator of the deformities: field biologists, like Dave Hoppe and Bob McKinnell, who understood exactly where frogs spent their days; toxicologists who knew how to test chemical pollutants that might harm early embryonic development in amphibians; and researchers who worked to uncover the basic processes of development, especially of limbs and eyes.

One priority was to analyze the pond environments and frog tissues for the presence of pollutants and other possible causative agents. Each identified agent would then have to be tested in a lab to see if it could cause deformities in developing frog embryos.

One day I grabbed a pink message slip from my mail slot at the office. An EPA scientist, Roy Folmar, had called from Florida. "He may have the answer to the deformed frogs," noted the support staffer who'd taken the call.

Recently, Roy had found feminization in male fish in the Mississippi River below the sewage treatment plant in St. Paul. He was actively researching chemical pollutants that disrupt the actions of normal hormones such as the sex hormones estrogen or testosterone. Endocrine-disrupting chemicals were getting a lot of attention in the media because of their dramatic effects on reproduction. They could deform penises in alligators or cause female ovarian tissue to grow in the testes of males. Could some of these disrupting chemicals also cause limb defects in frogs? I knew that a wide array of chemicals released into wastewater and rivers from detergents, plastics, certain drugs, and pesticides could disrupt hormonal systems in animals.

I called Roy, and we chatted about his fish work. "The boy fish were looking like girl fish," he drawled. Some pollutant in the water discharged from the sewage treatment plant had acted like the hormone estrogen and caused the male fish to produce an egg-yolk protein normally made only by females. Later, feminized male fish were found swimming in water not only downstream but upstream of wastewater treatment plants in rural areas. This finding suggested that water running off from farmland also contained chemicals that disrupted the hormone balance in male fish and caused them to make the yolk protein (Lee et al. 2000). More recent work in Minnesota has revealed several hormonally disruptive contaminants, in lakes as well as rivers, including chemicals from plastics and detergents, hormones likely from human origins, certain drugs, and triclosan, an antibacterial agent in household products (Ferrey et al. 2010).

"But that's not what I called you about," Roy said, shifting gears. "Do you know how similar thyroid hormone and DDT are?"

"Not really," I replied, wondering where he was headed.

Roy's voice rose in pitch. Excitedly, he tried to describe in words the similarities in the chemical structures of thyroid hormone and DDT, the long-banned insecticide. I perked up. Thyroid hormone (thyroxin) is a key chemical in frog development. It drives the metamorphic transformation from tadpole to adult. What if DDT or some other pollutant could mimic thyroid hormone and cause harm to developing frogs? Although banned from use in the United States in 1972, DDT continued to persist in the environment and still shows up in tissues of some wildlife and humans.

Structurally, they each had two aromatic rings made of carbon with halogen atoms attached (chlorine in DDT, iodine in thyroid hormone). But were they similar enough in chemical activity to interfere or compete with each other in a living organism like a frog?

Evidence was lacking that DDT could mimic thyroid action, I said, hoping Roy might know otherwise. Nor was I aware of any other potential disruptors of thyroid hormone.

"You're right, he said. "But it's one way deformities might happen. You should think about it."

DDT went on our list of suspects along with any possible chemicals that might interfere with the action of thyroid hormone. Several years later, EPA would develop tests to screen chemicals for activity that disrupted normal thyroid hormone action (see epilogue). By then, chemicals such as 2,4-D, DDD (a breakdown product of DDT), PCBs, and dioxins would be shown to interfere with thyroid hormone. In the next decade, concerns would be raised that nitrates from farm fertilizers might impair thyroid hormone function and possibly be linked to thyroid cancer and slowed development in frogs (see epilogue).

A steep road lay in front of us, one paved with many unknowns. I forged ahead, posing questions to scientists who had greater expertise than I, my habitual way of learning, As I picked up new information, I would pass it on to anyone who wanted to listen.

As part of my quest, I explored the existing scientific literature. What had previous researchers uncovered about specific pollutants or biological agents that could harm frogs? On weekends I accessed the online databases at the university library, where I tried various search terms, such as "malformations and frogs," to find any relevant research that might give us clues to causes of the deformities. I obtained a draft copy of a report by a Canadian scientist, Bruce Pauli, who had assembled a bibliography of journal citations related to amphibians and pollution (Pauli et al. 2000).

But to my dismay, research on specific pollutants that might harm frog development was slim. Most laboratory tests for toxicity of chemicals in water were run for only a few days, and these tests tended to focus on acute mortality—how fast do how many critters die? Some experiments measure effects on egg production using invertebrates that have short life cycles, like *Daphnia,* which lives at most a few weeks.

Rarely were developmental defects studied in vertebrates because such tests were costly. However, one experimental test involving exposure of young embryos of the African clawed frog (*Xenopus*) did measure very early developmental abnormalities. Designed to provide a rapid screening of chemicals, in most cases, this test was not carried out long enough to allow frog larvae to sprout limbs, let alone long enough for juveniles to transform into adults. The *Xenopus* test was criticized because the biology of the African clawed frog differs markedly from native frog species.

Later, our research partners would use the standardized *Xenopus* embryo test on pond water taken from our study sites.

Pat Schoff, a developmental biology researcher at the Natural Resources Research Institute in Duluth, called. His interest was retinoic acid, an important natural molecule related to vitamin A that directs early development. Retinoic acid (or RA for short) is essential in the formation of structures like legs and internal organs. It, or molecules imitating it, could cause limb deformities, he said. I knew that pregnant women were warned not to use an acne cream that contained retinoic acid, because it could cause birth defects.

He faxed me a research article that had photographs of abnormalities that developed in chickens, mice, and frogs caused by excess retinoic acid. I stared at the fuzzy images of frogs with weirdly shaped or multiple limbs and felt a chill of recognition. The deformities looked very similar to some we'd just seen at the Ney Pond.

I'd placed parasites on my early list of potential culprits because of a scientific article published in 1990 (Sessions and Ruth 1990). The authors had been surprised when they found multilegged tree frogs and salamanders in a pond in northern California. During internal examinations of the frogs, they observed rounded resting stages, called cysts, located near the base of the limbs. The researchers concluded that an immature stage of a parasitic trematode flatworm had caused the multiple limbs. Their hypothesis suggested that the cyst stages, which tended to embed near the base of early developing limbs, could block the normal chemical signals that control the outgrowth and proper development of the limb.

Parasites as a possible cause of deformed frogs presented a different kind of puzzle, a prickly one. For one thing, parasites have been cohabiting and coevolving with frogs for millennia. Frogs are hosts for many kinds of parasites, like flatworms and roundworms, which live inside their gut, bile duct, lungs, and kidneys. Why would a particular parasite suddenly start causing developmental defects in frogs during the 1990s in more than one species and in many locations in the United States and Canada?

Scientists I talked with agreed that we needed to determine whether parasites caused any of the deformities. But at the same time, most of them felt that chemical pollutants were far more likely. The majority of the malformed frogs that people saw had missing or partial limbs; only a small fraction of them had multiple or branched limbs, which some scientists did attribute to parasites.

Because many of the deformed frog sites were located in farming areas, agricultural pesticides remained high on the growing list of chemical suspects.

A Canadian researcher, Martin Ouellett, had openly targeted pesticides as a likely cause of the malformed frogs he'd found in Ontario during the early and mid-1990s. In his extensive frog surveys, he found more deformed frogs near corn and potato fields than he did in nonagricultural areas.

"Give out a lot of fog about agricultural chemicals" to the public, my section manager advised me one day. Stunned, I took this to mean that he felt I shouldn't say we included farm chemicals among our list of other possible causes of the deformities. Did this come from the MPCA's higher level managers? Were they pressured by the state's Department of Agriculture? I responded that we had to say honestly that pesticides are on our list of suspects. How would my agency react if we discovered that a particular farm chemical had caused the deformities? I wondered.

Pesticides are a political hot potato in a rural state that has strong economic ties to agriculture. To my surprise, I'd learned earlier that Minnesota's Department of Agriculture, not the Pollution Control Agency, had sole legal and regulatory authority over pesticides. Yet the MPCA protected water quality in the state. I had wrongly assumed that my agency, which oversaw the state's water quality, would regulate all pollutants, pesticides included.

To protect state waters, the MPCA develops specific rules and standards for regulating pollutants that are discharged into streams and other water bodies. The agency has many standards in place for toxic chemicals, such as mercury, PCBs, and ammonia. Only a few pesticides are in the MPCA's legal rules, and most of these are chemicals, such as DDT and chlordane, long banned from use. More recently, the state's rules list a mere thirteen pesticides with water quality standards for surface waters such as rivers and lakes. Only twenty-four have standards aimed at protecting drinking water (Minnesota Administrative Rules, Chapter 7050 and Chapter 4717). Hundreds of farm chemicals currently used across the state still have no regulatory standards, not at the MPCA, which could develop them, and not at the state's Department of Agriculture, which has the legal authority over pesticides.

Neither the Pollution Control Agency nor the Minnesota Department of Agriculture have the standards in place that would provide the scientific basis for interpreting environmental data for most pesticides. It's a regulatory conundrum.

One day a man called who wouldn't identify himself. Later I referred to him as my "Deep Throat," because of his insistence on anonymity and his deep, serious way of speaking.

"Have you looked into the herbicide acetochlor?" he asked. "It's new on the market and there's pressure to increase its use." Acetochlor had been given what's called a "conditional use" registration by the EPA in 1994, he said, because EPA wanted farmers to reduce the use of other major herbicides, like atrazine and alachlor. Acetochlor had side-stepped the EPA's formal process; its use was allowed without review.

"It looks like Minnesota farmers are using acetochlor in much greater quantities than in other states," he said. How could he know this? I was dying to find out who he was and where he worked, but he wouldn't say. I told him there were also deformed frogs in some parts of Wisconsin, and in northern Vermont.

"Which counties?" he asked. I grabbed a memo sent to me from a DNR biologist in Wisconsin; as I named the counties, I could hear his computer keys clicking as he read off information on acetochlor use.

I had so many questions I wanted to ask him, but his voice tightened and he signed off. Who was this man? How did he have such ready access to information I couldn't possibly get my hands on?

He never called again.

One scientist in the state's Department of Agriculture shared information with me about pesticides that he had recently monitored in southeastern Minnesota. He had found high levels of certain pesticides in stream water in May and June in his surveys, surges that followed the usual heavy applications of chemicals to crops in spring. Snow melt and spring rains flush them into streams and, for that matter, into wetlands where frogs deposit their eggs.

Acetochlor was already showing up in stream water by the spring of 1995, he said, only one year after the EPA had given the chemical a conditional use registration. My anonymous caller was right about this new herbicide. Minnesota farmers were rapidly shifting over to acetochlor; by 2005 it was being used on one third of the state's corn, and by 2009 the EPA would list acetochlor as "a likely human carcinogen" (EPA 2009a, 2009b).

Perry Jones, a scientist located at the local United States Geological Survey (USGS) office told me they had found fairly high amounts of farm chemicals coming down in rain, especially in spring, and even in the Minneapolis–St. Paul area (Capel et al. 1998). Rainfall! The distilled liquid I thought was relatively pure. The water some rural people collected for drinking water could be contaminated by airborne pesticides.

Not only that, a study by the USGS of wells in twelve midwestern states found mostly low levels of herbicides in 40 percent of 303 wells tested (USGS 1998). More recently, pesticide exposure in agricultural areas and in rural drinking water has been linked to the development of Parkinson's disease

(Gatto et al. 2009). Disturbing news, but did this have anything to do with the deformed frogs? At best, these studies showed correlations, not cause-and-effect relationships.

What I did know was this: spring is the critical time for frog reproduction and a likely time of exposure to chemicals running off the land into the water. It is the time when eggs are laid, the time when the earliest embryonic stages are exposed to whatever is in the water, chemical or biological. This was the most important window of time for us to get out and sample the frog ponds.

In 1996 an enthusiastic USGS chemist from Kansas, Mike Thurman, accompanied us in the field in late spring to sample a few frog ponds for farm chemicals and other organics. Mike was especially interested in looking for methoprene, an insecticide used to control mosquitoes and shown in lab tests to cause development defects. I knew methoprene was widely applied to shallow wetlands around Minneapolis–St. Paul to fight mosquitoes. It also decimates other kinds of aquatic invertebrates, and I'd long spoken against its use in wetlands for this reason. But methoprene use was limited in rural Minnesota, where many deformed frog reports originated.

As we drove to our study sites, Mike talked about the fate of pesticides once they get into the environment. Microbes and other factors break down the active ingredient of a pesticide, the part that kills weeds or insects, in a sequence from one molecule to the next, he said. These breakdown products may persist in water or soil, and no one knows if they're toxic because they are not tested (Koplin et al. 1998; Thurman et al. 1992).

"We don't even know whether any of the thousands of so-called inert ingredients that are added to the active ingredient are safe," Mike said, explaining that the identities of inerts are protected as trade secrets. The EPA does not require their disclosure to the public, let alone require toxicity testing on them, Mike said. Even today, this remains an issue of concern.

I needed to find out which pesticides were used near the frog ponds. But the more I talked with farmers, the more I realized that information about pesticide use was going to be hard to come by, if not impossible. Some farmers were willing to talk with me, but they had mixed records of what they used. They applied different chemicals on different crops in different years. Don Ney identified three herbicides he used near the Ney Pond, but his neighbor, whose extensive crop fields drained runoff in tiled pipes that ran into the Ney Pond, would not tell me what chemicals he used. Much later, USGS scientists would identify several pesticides coming into the pond through those pipes (Jones et al. 1999).

When farmers use a pesticide that is designated "restricted use" by the EPA, they have to pay licensed professional pesticide applicators to douse their fields. The registered applicators have to submit records about which pesticides they used and where to the state's Department of Agriculture. But to my great frustration, this was not (and still is not) public information. It is sequestered by the department. When I asked for specifics, I was told the agriculture agency considers pesticide use information *proprietary.* The agency appeared to be protecting the sales information of the agricultural chemical companies. Although I was hardly a competitor, I was denied access to their records.

A Minnesota citizen whose land has been oversprayed by a neighboring farmer, has no right to know which chemicals were used. Only if someone develops certain health problems can the information on local pesticide use be released—and then only to a doctor.

This was crazy: even if we could uncover which pesticides were used around the different frog ponds, we'd still be woefully lacking important information, such as the potential for developmental toxicity of each formulation's additives, the so-called inert ingredients, or for each pesticide's longer-lived breakdown products.

Mark and I planned to analyze pesticides in water samples from a small number of frog ponds during the spring of 1996. We'd collect the water at the time the frogs would be laying their eggs. The Minnesota Department of Agriculture had analyzed a few samples of pond water and sediments taken from the Ney Pond in the fall of 1995. No pesticides were at "levels of concern," they'd said. I didn't know what that meant, since they had no specific chemical standards, no scientifically defined "levels of concern." Moreover, their collections took place late in the fall, not during the crucial spring season.

For the 1996 work, we decided to have pesticides analyzed by an independent lab, if only to avoid the appearance of any possible bias toward agricultural chemical industries if we used the state's agriculture lab. Mark called the scientist who'd done the analysis to ask for a copy of their field methods for sampling water and mud. We wanted our collecting protocols to be consistent with those used by the agriculture lab.

A simple request for information, we thought.

This time, the scientist who had helped us had to inform his supervisor about our request before he could even send us the sampling protocols. Our request traveled up the management chain to the Department of Agriculture's commissioner, who was, apparently, meeting with reps from agricultural chemical companies at the time. Someone at that meeting purportedly said that it sounded like the pollution control agency was on a "witch hunt."

This comment passed from the agriculture commissioner to our agency's commissioner, who then passed it down to Mark and me. We sat stunned when we heard this. *Witch hunt?* Apparently, someone did not want us measuring pesticides at deformed frog ponds.

One of my bosses had warned me that agricultural chemical companies would spend hundreds of thousands of dollars to discredit a small study that criticized pesticide safety. If a pesticide or its breakdown products or any its additives were responsible for the deformed frogs, Mark and I knew that the chemical analysis on frog ponds would have to be first rate, impossible to discredit.

How naive I had been to think that everyone, including people in public agencies, would want the truth about the frogs. My learning curve was encompassing far more than science. The path ahead was tangled. The stakes were high. But at this point we would not turn back.

"There will be no funding for the frogs!" our section manager said, as he slammed down a stack of files on the conference table before starting a grueling meeting to discuss Mark's and my work progress. Was he angry because the upper echelon of the agency told him to say this? His emphatic, angry tone rang in my ears. His outburst reminded me of my dad's temper flares, often related to his personal comforts and always directed toward my mother, who seemingly tolerated them. Once she told me he was just being grouchy, but I didn't see it that way.

Before this meeting, Mark had been worried. "We've been eating up our federal wetlands grants to pay our salaries to work on the frogs," he'd said, looking grim and wondering how long we could keep this up. Our grant-funded positions were, essentially, still temporary. We faced a tough dilemma: if we continued consuming our EPA money to work on the frogs, our wetlands work would suffer. Could we sustain two major research projects, one to continue our main work developing a wetlands-rating system using measures of the aquatic invertebrates and wetlands plants, the other researching causes of deformed frogs? If we pushed the agency to fund the portion of our work time involved in investigating frogs, might we lose control of our wetlands work? Might we risk our jobs?

As 1995 had drawn to a close, any possibilities for funding additional work on the frogs for the 1996 field season that lay ahead looked nonexistent. Absent new resources, we couldn't hire seasonal field crews or pay for technical analysis. In addition, I was dreaming of a program to reach out to more schoolchildren across the state, kids like the students who'd discovered the frogs at the Ney Pond. They could be the "eyes" for us—go out and hunt for frogs near their schools and report back to us what they saw. I even had a tentative title for a school-based effort: "A Thousand Friends of Frogs."

Children have different ways of seeing nature: they're better at it than adults. At least they proved this at the Ney Pond in 1995. Adults might have walked blindly past the deformed frogs. The same was true for the quest for migrating frogs that fall. Students spotted frogs hiding in crevices in the river bank; they found them under water in spring-fed flowages in the river basin.

When my sons were preschoolers we were lucky to live in Decorah, Iowa, on a lovely wooded hill, a place where they could roam freely with neighborhood kids. Together this little pack explored the woods and fields, built forts, and found natural treasures. A dry-run, stony streambed ran down the wooded hill close by, and there the boys discovered stones shaped like three-inch-long snails and clamlike creatures. They brought home pieces of flattened limestone slabs bumpy with the prints of ancient molluscs, sponges, and crustaceans. We looked them up: those fossils were at least a hundred million years old, and the hill where we lived had once been covered by an inland sea! One time my younger son Steve, barely four, picked up a small brachiopod fossil that lay hidden in a gravelly path, an exquisite, clamlike animal that I'd have missed completely. Kids *are* closer to the earth. One fall Erik had captured a tiger salamander, which we kept in a ten-gallon terrarium and fed hamburger. In winter it buried itself in the soil; in spring it miraculously pushed up out of the dry soil just when we feared it had died, especially since I'd neglected to water the terrarium. Its skin was slimy, and it looked remarkably robust after some months underground.

Today the importance of getting children outside to explore and manipulate their environment on their own has gained greater credence. Children are becoming more and more isolated from nature by all sorts of electronic media; they are less likely to roam freely because of parental fears. If outdoors, they're running around on sterile, green ballfields. Studies suggest the consequences of this cultural change are serious. Today's children are more likely to suffer attention deficit problems and other disorders. In his 2008 book, *Last Child in the Woods,* Richard Louv lays out the urgent need to let our children explore outdoors, partly because this can ease attention deficit disorders in them. Our children have, in Louv's view, a condition he calls "nature deficit disorder."

The idea of funding a program to prompt teachers and students to get out and look for frogs appealed strongly to Representative Munger. He'd reached out that fall to the Minnesota New Country School students, inviting them to speak at his hearing on frogs and wetlands. Munger's bill for frog research included money to educate schoolchildren.

Munger was our only hope. He was trying to force the MPCA to embrace the frog problem; he'd indicated the money should go under my management.

A conflict arose. Our upper level manager had previously told me firmly that the MPCA *does not* gets its funding from proposals initiated by legislators, Munger or otherwise. The agency's budget has to go through the governor before it gets to the legislature for review.

This reality put us on a collision course. As Representative Munger prepared his bill to provide MPCA money to fund an investigation into the deformed frogs, the MPCA stood ready to block it.

Gail Thovson, the science teacher whose student had found deformed frogs that fall at a rural pond north of Litchfield, Minnesota, organized her science students to write to the governor and implore him to fund work on the frogs. Science teacher Thovson said her students were always asking "How can we help? How can we get some answers?"

"I told them it takes money and leadership," she said to me later. She'd read in the newspaper that the MPCA claimed it had shortages of funds for work on the frogs. Yet she knew the state had a surplus of money and I needed funding for an investigation.

The letter from Thovson's middle school science students addressed to Governor Arne Carlson arrived at the governor's office on January 2, 1996. The topic: "Funds for continuing the deformed frog investigation."

"We care!" the students' letter said, "for us, for our future, for your children's future." They asked the governor to set aside the money necessary to complete an emergency investigation of the frogs. Attached to the letter were several sheets of paper with signatures from students in four science classes along with their hand-written pleas. Attached to the letter were statements from Thovson's own students.

For instance, a student named Holly wrote: "Why did this happen? Why are these frogs deformed? Were the frogs [eggs] laid in some kind of pollution? . . . We need to find out . . . These frogs are deformed because one has no legs. One has one leg that's not all the way developed. One of them has something wrong with its jaw so it can't eat very well."

Another, Lisa, said: "This frog business is really making me think . . . Think! Even people could eventually be turning out this way."

And Rebecca wrote: "I think this happened because of the environmental changes in the Litchfield area. The owner of the land where the frogs were found said they had not been able to drink their well water for about ten years. An unusual number of cancer deaths have occurred in that area. . . . If we learn more about this problem, maybe we can prevent it in other places."

Their letter of petition landed on my desk in a transparent yellow plastic folder with an official agency routing slip that indicated it had come from the governor's office. It felt surreal that I was expected to write a response explaining the funding situation to the students, a response that would go under the MPCA's commissioner's name, not mine. I happily praised the teacher and her students for all they had done to alert us to the deformed frogs near Litchfield. Their finding of deformed frogs in the fall of 1995 had expanded the significance of the problem into another area of Minnesota.

After my sentences, MPCA management added: "We are not always financially able to respond to new problems that arise unpredictably during the biennium [the two year state budget period]. The MPCA is interested in receiving funding to continue the frog investigation *only if the additional funding will not be removed from its current priority budget requests.*"

Representative Munger asked me to make a presentation at a legislative hearing to be held at the end of January 1996 at the State Capitol. Its white-domed, granite building sits on a hill facing downtown St. Paul. On the dome's rim stands a golden sculpture, heroic-looking human figures who seem to be struggling to restrain several muscular horses from leaping off the edge. The halls of the State Capitol are lined with portraits of past officials, statuary, and Indian flags. They frequently echo loudly with the footsteps and voices of visiting students, demonstrators, legislators, and tourists.

I walked up the long flight of marble stairs and entered the hearing room, solid-looking with its heavy, dark woodwork and tall windows. Legislators were standing around talking, about to sit down at a long table lined with name plates and microphones. Up near the front I spotted Bob McKinnell, who had come at my request: I knew Bob, a university professor, would have more prestige in the eyes of legislators than I had as a mere state employee, especially one from the MPCA.

When called up to speak, I described what we'd learned so far about the deformed frogs and what we hoped to do next. I distributed several color photos that I'd taken of limbless and extra-legged deformed frogs at the Ney Pond and mounted on pieces of foam core board. Legislators murmured as they passed the photos around the table. In this case, one picture said more than all my words could ever convey.

Speaking with his usual energy, Bob McKinnell conveyed a sense of urgency about the extraordinarily malformed frogs he'd seen the past summer.

He pulled out his wallet, opening it wide to dramatize that he'd been using his own money to work on the frogs.

As the hearing concluded, the chair of the finance committee came up and asked, "Dr. Helgen, can you and Dr. McKinnell come out in the hall to talk with Representative Munger?" We huddled in a cluster in the spacious hall. The finance chair asked me: "Could you assemble a budget for $150,000 for frog research and include in it a budget for participation with school students and get it to me by Monday?"

I gulped and said I would. It was already Friday afternoon; for me, a long weekend at the office lay ahead. Verlyn was out of town for a family funeral; my time was relatively free. I'd have to work fast and design a thoughtful budget with no input from my bosses, who likely would have stopped it cold. Fortunately, I already had a good idea about which elements to include in a proposal.

Then a medical researcher, a man I'd never met but had seen sitting beside McKinnell at the hearing, came over and stood with our small group as we talked in the hallway. I wondered why he was there. Unabashedly, he asked Munger for *half* of the $150,000 to use for his own research related to humans and pesticides. I stiffened. How could he co-opt money aimed at frog research? He had never talked to me about cooperating or working on the frogs, so his audacious money grab came as a total surprise. I'd never seen anything like it. Later I heard he'd said to someone, "I hope this woman isn't greedy." Greedy? If anything, I'd serve only as a conduit for the money to flow out to other researchers and to teachers and students, not to me!

I suggested to him that it would be more appropriate if he submitted a separate proposal, since this money was aimed for frog field research and for starting a program with schoolchildren. "How can you tolerate working with all those bureaucrats?" he asked me. I stared at him blankly. Then he saw Representative Munger leaving and walked away beside him, talking closely as they moved down the hall. I could only pray that he would not get Munger on his side. He didn't. Munger stood firm. He intended that the money be used for work on the frogs, not a medical study.

Late that same day the researcher called me at work and told me with some vehemence that he was going to contact all his friends in the legislature over the weekend. He'd get his money from Munger's proposal, he said. Clearly, he played hardball. Would he torpedo the budget I was composing for the finance chair?

This researcher was examining the health effects of pesticides on humans. He observed higher rates of birth defects in the children of pesticide applicators in Minnesota, more chromosomal abnormalities, and higher incidences of lymphoma. He obviously believed strongly in the value of his research, which was, unquestionably, important work. And years later his concerns would be borne out by other scientists who would publish articles showing seasonal associations between pesticides, the timing of conception, and birth defects in humans (see epilogue).

The teacher Cindy Reinitz and I reflected about some of the people we had encountered because of the deformed frogs. She was still under threat and now having to attend hearings because of the fallacious charge made against her by an upset parent, a story she'd told me earlier. Hearing about my tangle with the medical researcher, Cindy commented, "You know, this problem with the deformed frogs brings out *the best and the worst in people.*" In the future, we had occasion to repeat this.

In a rush, I composed a budget that allocated money for Hamline University to involve schools in a new "Thousand Friends of Frogs" program and directed funds to four researchers, Bob McKinnell included, for work on frog chromosomes, histopathology and necropsy of frogs, and analysis of frog DNA. Experiments rearing frog eggs in the lab and in a pond that had deformed frogs would be done by Dave Hoppe. Some of the funding would go to the MPCA so we could form contracts for chemical analysis of frog tissues, pond water, and sediment samples. Some we'd use to hire temporary summer workers for field surveys of frogs. I e-mailed my bosses about what was going on, and early Monday morning I delivered the proposal to the legislator.

I didn't care much what my bosses thought at that point. I had to respond to the finance chair's request. I was not "lobbying," and without this money the frog investigation would die.

It was just as well that Verlyn was in South Dakota all weekend to conduct his cousin's funeral and help his elderly mother and her equally elderly twin sister. It freed me from one source of tension as I spent the weekend at the office. Our future path remained somewhat ambivalent, our dance continued. He'd drop hints about other women who'd expressed interest in seeing him (they had), and kidded about needing a "rich widow." His jokes tapped into my insecurities over relationships with men. Would I be able to tackle a marriage with all that was required to hold it together? Would he? We'd had long conversations about fidelity, aging, and our need for having some degree of separation, some independence. I'd known one close couple that behaved as if their identities were merged into one. Never apart, their degree of togetherness harkened back to the expectations of the 1950s and gave me a chill. Might I lose myself in a marriage even though Verlyn and I had agreed that having our own interests was healthy and necessary? He wouldn't try to control me, I knew that. And I doubted he could in any regard.

Within a few days Cindy Reinitz, her students, and I were assembled in front of a formal legislative committee for a hearing on Representative Munger's bill to fund frog work through the MPCA. Just before I left work to testify, I was told firmly by my division director, "If you are asked if MPCA supports Munger's bill, you will have to say no, because that's not the way we do our funding."

I responded immediately: "Then what do I tell all the students from the New Country School who are coming up here from Le Sueur to testify?"

"Judy shouldn't be put in that position," my manager had said to her higher-ups. The agency sent its assistant commissioner, who rushed over to the hearing to respond, but the question about the MPCA's nonsupport of the bill was never asked.

In a crowded hearing room with legislators seated around tables and reporters and others filling the chairs on the sides, I stepped forward to a small central table and spoke into a microphone. I gave an update about the frogs and showed slides of students in the field at the Ney Pond along with pictures of deformed frogs. I outlined our future plans, should we receive funding.

Cindy Reinitz eloquently summarized the school's discovery and her concerns about the frogs and the health of the environment; then she introduced her students, who came up to the microphones at the presenter's table and spoke directly to the legislators.

At that point, any distracted legislators came to attention, and all eyes were riveted on these serious young people who pleaded with passion for someone to investigate the frogs they'd seen. Below are some excerpts:

Student Jack Bove began, speaking with a youthful energy: "When someone caught a frog with only one leg I thought to myself, 'Houston, we have a problem. . . . What's going on here? Can whatever have deformed these frogs affect us?"

Next, a tall and stoic-looking teenager, Nick Pollack, spoke, his voice strong beyond his years. People listened: "We must find a reason why the frogs are ending up deformed: the frogs are a measure of how the ecosystem is doing . . . We need to find the solution *fast.*"

Ryan Fisher, looking solemn, addressed the assembled legislators: "Just because we don't know about any effects on humans doesn't mean there aren't any. We just don't know! . . . I've always been wary of the environmental disasters and dangers that some people believe in. . . . But with the frogs, we already have a problem."

Reta Bove, who had loved frogs from childhood, spoke with a quiet passion: "We need to listen to the frogs so the problems don't get worse so I can grow up in a healthy world. . . . I wonder what life will be like for my kids. . . . will they even be normal? . . . Our frog group is an example that kids can do a lot."

Finally, Betsy Kroon, who later helped produce an award-winning video about the students' work on the frogs, testified:

> We know that scientists consider frogs as an "indicator species." This means they are a species of animal that seems to be very sensitive to changes in the environment. They act as a warning system to humans, but only if we are observant.
>
> As a fifteen-year-old who loves the outdoors, I worry. I worry that we might not be seeing the warning signals out there. What will that leave for my generation when we are your age? I've read and heard stories about the way our state used to be. Lakes and streams were clean and blue, filled with fish and aquatic life. . . .
>
> The students from my school and I are here today to wave a warning flag in front of you. We are deeply committed to our frogs, our Ney Pond, and to our earth. This is an awesome task for kids. We need to know that the legislature hears our concerns, is devoted to the future of our future Earth, and is willing to make hard choices to lead the state in environmental action."

As the students finished, we all sat silent and awestruck, close to tears. I saw jaded legislators come alive as the fresh voices of these impassioned students filled the air. Incredibly, the students had written their own statements, which flowed directly from their hearts. Glad to be treated seriously when it came to the deformed frogs, they displayed a maturity and thoughtfulness well beyond their teenage years. I was deeply moved by the sincerity of their words, by their pleas for adults to care for the future of the environment, and by their courage to reach out and try to make a difference.

They succeeded. We received our funding, and the MPCA did not block it after all. Thanks to some students from a storefront charter school in Le Sueur and to the schoolchildren from Litchfield who had petitioned the governor, we could go ahead and investigate the deformed frogs during the coming summer and fall of 1996. With funding secured, I believed we were on our way.

Immediately after their discovery of deformed frogs at the Ney Pond, several students at the Minnesota New Country School had formed the Frog Group, with Cindy Reinitz serving as their teacher/adviser. For several years they would be intensely involved, giving presentations all over the country and speaking again to the state legislature; they conducted many interviews with the media; they surveyed the frogs and tested the pond water. A few of them ran an experiment one summer at the school for a university researcher. This required early risings and daily changes of water in aquaria as well as recording data about tadpoles. Later their work was acknowledged in a research paper (Loeffler et al. 2001). The Frog Group didn't mind that kids from the local high school referred to their charter school as Kermit High. They rather enjoyed it.

A decade later, some of the students reflected on how their experiences with the deformed frogs have influenced their lives. Reta Bove Lind, who loved collecting frogs at the Ney Pond, earned her college degree in recreation, parks and leisure studies. She returned to work on staff at Minnesota New Country School for a while and has worked at the Ney Nature Center and a history center in Henderson.

Ryan Fisher, who developed the Frog Group's website, had been homeschooled before going to the charter school. He went on to college, graduating *summa cum laude* with a degree in chemistry and math.

Becky Madison Pollack, now married with kids, still bubbles with enthusiasm. She had a huge passion for rocks and geology as a kid, she said. In college Becky majored in geography, with minors in geology and earth sciences. She now serves as the director at the Ney Nature Center and organizes its educational programs.

Becky's husband, Nick Pollack, was one of the original students who'd discovered the deformed frogs. Now a tall, strong man who works in construction, Nick helps restore the historic buildings at the Ney farm, serves on the board for the Ney Center, and has an avid interest in local history. What he did as part of the Frog Group at the MNCS made him more community involved and a more independent thinker with broader horizons, Nick said.

When reporters flooded the charter school for interviews about the deformed frogs, student Betsy Kroon loved talking with the media, even on camera. Betsy developed the video about the students and their work on the frogs that won an award for the Frog Group. "The science part was exciting, but it didn't give me the satisfaction that sharing the story and writing about it did," said Betsy, who went on to major in journalism and women's studies in college. Betsy now works as a freelance writer and editor.

Late in February of 1996, Verlyn came back to Minnesota for a few days from California Lutheran University, where he was serving as acting campus pastor and teaching religion for one semester. I staged a dinner for us, setting out a dozen candles around the house to create a romantic atmosphere. By then I realized how much of a calming influence Verlyn was in my life, an anchor and a link to a world where people laughed, kissed, discussed politics, worked for peace, and were open-minded about religion. He was not only a good teller of jokes and an antidote to the toxic environment at work, he brought a loving, human connection I deeply needed. At the same time, I knew he wanted me in his life, that was clear. He loved my laugh, he often said. After his troubled relationships in the past I think he was grateful for our steady, loving friendship, our shared interest in issues of peace and justice, in theater, and in our separate churches. I could see us as partners, as equals. We had some differences in our background (mine, New England Congregationalist; his, midwestern Lutheran) and careers, differences that already enriched our relationship. We would laugh when Verlyn told friends that his elderly mother asked him once, "Couldn't you find a girl of *our* faith?"

CHAPTER 4

WADING IN AND LISTENING

Before coming to the MPCA, I observed first-hand the vulnerability of wetland organisms to damaging pollutants when I participated with EPA scientists in a research project. Their goal was to field-test the impact of organophosphorus insecticides on the biota of natural wetlands. Would the results from these more complex environments line up with their laboratory toxicity tests, conducted on just a few species under simpler conditions? To sample the study sites, located in the Crow Hassan Reserve near the Twin Cities, I walked happily up a hill through a restored prairie area where bobolinks nested and tree frogs called from nearby woods. My task: to analyze the zooplankton, such as *Daphnia*, and other tiny crustaceans, before and after the wetlands were sprayed from a helicopter with insecticides. Another crew sampled the larger aquatic invertebrates, such as immature insects, snails, and clams, while EPA staff monitored the water chemistry, acidity, and daily swings in oxygen levels and temperature in the water. Our field results were stunning. The insecticides killed the zooplankton outright; the other invertebrates, like juvenile insects, declined over time.

I began to learn that the ecology of wetlands differs considerably from that of streams. Mayflies, certain midges, and caddis flies were already known to indicate good water quality in streams, but what about dragonflies, crustaceans, snails, leeches, beetles, bugs, and other invertebrates that are more abundant in wetlands? Oddly, some aquatic species that succeed in unpolluted wetlands, like certain types of leeches and midges, are considered pollution indicators in healthy streams. Stream-dwelling invertebrates depend on a steady supply in flowing water. Wetlands invertebrates are better adapted to survive big changes in the levels of dissolved oxygen in the water, which fluctuate daily with afternoon highs (oxygen released from photosynthesis by pond algae and plants) and overnight lows (oxygen removed after dark by microbial decomposers). Some, such as *Daphnia* and certain midge larvae, make a special red hemoglobin that helps them hold oxygen when it's low in the water. I'd seen red *Daphnia* in sewage ponds and at the Crow Hassan wetlands, indicating oxygen was low. So the day the EPA crew's test minnows went belly up in their floating cages, I wasn't surprised.

If forced to pick one group of species to indicate wetland water quality, I would probably suggest the presence of diverse, immature dragonflies. Ferocious underwater predators, juvenile dragonflies consume mosquito and midge larvae, small minnows, and many other prey. They are known to accumulate pollutants like lead, DDT, and toxic PCB oils from the organisms they eat (Corbet 1999). In Japan the exquisite, bug-eating adults are venerated as symbols of strength, courage, and happiness, perhaps because they zoom over rice paddies and are considered harbingers of rain. However, to develop a robust method for scoring water quality, whether in streams or wetlands, one needs to survey many different species of invertebrates, not just one "indicator" type.

My earlier work on zooplankton was conducted from docks the EPA had constructed in the three ponds we had studied. But when I started researching wetlands invertebrates at the MPCA in the early 1990s, I wanted to analyze the broader community of species. To do this, I used a long-handled dip net to sweep through the water and underwater vegetation in a repeatable manner. And I deployed underwater funnel traps overnight to capture the more active swimmers, with the help of intrepid assistants like Kyle Thompson and Cade Steffenson.

This meant wading into wetlands.

Early in my life, I wore high-top tennis shoes to fish for trout in rocky streams set in the balsam-scented, forested mountains of southern Vermont. I had little experience wading in mud-bottomed wetlands, shallow depressions that at best are fringed with diverse aquatic plants, at worst have shorelines packed with dense stands of cattails. At the MPCA, I started out using a clumsy pair of rubberized men's chest high waders that were so stiff I could barely raise a leg to step over anything. I stomped eagerly into a wetland, gear in hand to test my methods. Unexpectedly, an underwater log snagged my boot and I fell forward, face down in the water. My hands futilely sought leverage in the muck. Rotten odors surrounded my face. I wallowed, floundering in only two feet of water.

Smiling, Kyle, my curly-haired, robust six-foot assistant, a guy who knew wetlands well from his hunting and fishing experiences, waded over. Embarrassed, I reached for his hand and struggled to stand upright. "I see we should not sample wetlands alone, even in shallow water," I said, awed by my new-found vulnerability.

Soon I discovered much more flexible stocking-foot waders, the type used by fly fishermen, and slip-on, lace-up canvas boots. After that I hiked down to ponds and waded in without mishap. Only a few extremely polluted wetlands gave me trouble: one a former hog-feeding site that had muck so deep and tenacious I had to pull each leg out with both hands. Another was a large wetland that received enormous quantities of silt and polluted runoff from a major highway close to the Twin Cities. We were warned not to wade there because the bottom was like pudding at least ten feet down. To take water samples, I crawled cautiously over deeper water on a mobile, floating cattail mat, hoping I wouldn't slip through. If I did, I'd have to tread water and swim to shore.

Some of the more degraded wetlands we studied had junglelike stands of cattails, so dense they crowded out most other types of plants. Others were choked with masses of unwanted, submerged aquatic plants. Such excessive growth results from fertilizers swept in from farm or urban runoff. The worst ponds experienced heavy silting, chemical pollution, and biologically damaging fluctuations of water levels during storms. They looked like dead zones, barren of vegetation and almost devoid of invertebrate life. In one urban pond, we dipnetted a dead dog; in another, a decomposing loon.

The undisturbed wetlands, the ones we called "reference sites," were far more diverse and pleasant to sample. Their bottoms were firm underfoot, not steeped in soft urban silt or farm-generated muck. Mark taught me to marvel at the diversity of plants in healthy wetlands: the different kinds of sedges, grasses, and rushes; blue flag iris near shore; plants in the water with stalks bearing elegant white flowers; and my favorite, the yellow-blossoming bladderwort, an innocuous-looking, floating carnivorous plant that traps and digests tiny zooplankton and small mosquito larvae. I'd see its small blossoms projecting from the water's surface and envision the invertebrates disintegrating in the food-blackened bladders that floated beneath the surface.

At the high-quality wetlands, dragonflies zoomed around, frogs called, and our dip nets yielded a wide variety of invertebrate life that cruised and crawled about in the sample trays before we sieved and pickled them to analyze later in the lab. The brilliant, iridescent colors and patterns of adult dragonflies—metallic golds, reds, and intense blues set against blacks and greens—and their aerial antics as they captured mosquitoes on the fly enthralled me. In sharp contrast, the inch-long, aquatic juvenile stages of dragonflies have dull-brown bodies armored with plates and spiny projections. They look like miniature prehistoric monsters, yet they are very sensitive to pollution.

Once I stood in thigh-deep water and watched a dragonfly larva that had crawled out of the water onto an emergent plant stem, which it had gripped tightly with its spindly legs. Slowly the outer shell split open along the midline of its back. Then a compressed dragonfly pushed itself out of the shell, leaned briefly backward, then righted itself, unfurled its wings, and began transforming into the elegant adult.

At another unpolluted wetland, the pond's surface looked gray as we approached it on a blue-sky day. I waded in, surprised to see thousands of dead, delicate mayflies floating on the surface. Earlier, they had emerged from the water as mouthless, winged adults whose only goal was to mate. The females then returned to lay their eggs on the pond's surface and die. Rarely did polluted wetlands produce mayflies, but here they abounded.

In the early 1990s, the EPA assembled a national working group of wetlands scientists to research various approaches and develop biological indicators appropriate for evaluating wetlands (the Biological Assessment Working Group for Wetlands, or BAWWG). The EPA was promoting the use of biological methods to assess the water quality of wetlands and invited Mark and me to join the group. The first task was to research which species are sensitive to polluted water and which ones tolerate it, then develop methods for rating the pollution status or condition of wetlands. Various categories of plants and animals, such as algae, plants, invertebrates, and amphibians, were researched by several scientists who participated in BAWWG (US EPA Wetland Bioassessment Publications).

The work of the BAWWG researchers tapped into a wider, national movement supported by the EPA to encourage states to use biological monitoring as a more integrated framework for monitoring pollution and gauging water quality. The EPA had earlier encouraged states to do this for rivers and streams, using changes in fish communities to assess water quality in flowing waters. Ohio had pioneered this approach, and the MPCA was just beginning to evaluate fish in rivers as pollution indicators.

In the 1990s the EPA added wetlands to its biological monitoring wish list for states. But would the MPCA ever agree to monitor wetlands using our new biological methods? Convincing our bosses to do that seemed daunting. For us to use federal funds to develop the scientific tools to monitor the health of wetlands was one thing; having state leadership with the political will to implement these methods and do something with the results, like working to improve the quality of wetlands, was another. Once, in a discussion about polluted storm water runoff, an MPCA manager had commented offhandedly, "Well, you've got to put it somewhere," as if wetlands served only to protect lakes and streams.

Over the years Mark and I would refine our biological rating systems for wetlands (Indexes of Biological Integrity, or IBIs), using the methods we had standardized based upon our field research in dozens of wetlands. Our data showed that high levels of pollutants, like phosphorus from fertilizer runoff and chloride from road salt and farm chemicals, correlated significantly with losses of both invertebrate and plant species.

In 1996 Mark and I were launching a new project, supported by the EPA, to train volunteers to monitor the biological health of local wetlands. We hoped to foster a greater awareness not only of biological diversity in clean wetlands but also of the need for regulatory protections of water quality in these neglected habitats. Could informed and inspired volunteers learn to care enough for local wetlands to try and protect them? To ask government to stop using wetlands as the place to run polluted water from highways and mall parking lots?

The lower status and lax protections for wetlands made it easier for developers and cities to justify taking out a wetland to make way for more "worthy" projects. Those "useless" wet areas could bring in more tax revenue if they were converted to strip malls or residential developments. And wetlands bred mosquitoes; they were unsafe places for kids. Better to let them collect storm water or drain and fill them to support profitable properties.

Mark and I believed in getting volunteers outdoors and into wetlands, where they could discover and observe the diversity of tiny animals and aquatic plants that thrived in healthy, unpolluted wetlands. Wading in and looking at wetlands life might more readily change attitudes than hearing us lecture at them indoors. Of course we wanted to teach the volunteers about wetlands biodiversity, about the thousands of kinds of aquatic invertebrates that inhabit different types of freshwater wetlands and the nearly two thousand species of birds that are wetlands dependent. Knowing these facts and seeing wetlands diversity firsthand are two different things.

We planned to teach teams of citizens, led by high school biology teachers and often including local government staff, from several communities to sample wetlands invertebrates in June and the aquatic plants in July. Mark and I each streamlined our methods using invertebrates and water-dependent plants to rate wetlands so the volunteers could make a broad-brush determination of the water quality of their wetlands. With funding from the EPA, we partnered with Minnesota Audubon to help organize the teams and provide logistics for the training sessions that Mark and I would lead.

But in the spring of 1996, just before the start-up of this new project, a political uprising almost killed the new Wetlands Health Evaluation Project. In its newsletter, Audubon had announced our plans to initiate a volunteer monitoring effort for wetlands in partnership with the MPCA. Shortly thereafter, a flood of complaints poured into the MPCA commissioner's office, castigating the project. One man was angry that the MPCA had partnered with Audubon and its "radical agenda." Another asked why the government needed to find out what plants and creatures lived in wetlands. Might we find some unusual species that could cause Minnesotans to lose their jobs, like the spotted owl out west? One asked why we were creating a "cadre of citizen-spies to report on fellow citizens." He referred to Audubon as a "GAG" group, or "green advocacy group," and asked why tax dollars were funding a "special interest wetland vigilante group."

Vigilantes? I imagined creating "Wetland Vigilante" tee shirts to give to our radical volunteers.

This conflict, seemingly orchestrated, tapped into a larger issue that simmered in parts of Minnesota, the issue of individual property rights. A landowners' rights organization was advocating the view that government should not control what people can or can't do on their land. Under the banner of "Wise Use," a loose coalition of groups was fighting against government regulations over both private and public land. If government restricts property rights, they argued, then government should compensate owners financially for any losses they might incur.

Many of Minnesota's wetlands sit on private land, which makes regulating them politically contentious. Controls over activities like draining, dredging, or pollution met strong resistance. Some individuals had purchased cheap land containing wetlands that they planned to drain and resell for a profit. Angry, they butted heads with government and expected to be paid *not* to drain their wetlands.

Did this belief mean that individuals have the right to pollute their own land? I wondered: What if such a belief in individual property rights had prevailed during the earlier decades of major environmental cleanups? We'd have people dying of lead and dioxin poisoning, fish unable to survive in many rivers, a sterile Lake Erie, and more rampant loss of wetlands.

Someone asked Mark if training citizens to evaluate wetlands would give Audubon an "inside track" when lobbying the legislature on wetlands issues. Mark replied wryly, "Our information is available to anyone . . . if knowing how to identify wetland invertebrates, plants, and frog calls is having an inside track, then perhaps you're right."

Because of the political storm stirred up by our proposed monitoring program, the MPCA's commissioner attempted to shut it down for a year. But Mark and I fought desperately and saved the project. Over the next several years the Wetland Health Evaluation Project became very popular and involved a dozen or more communities. I was pleased when it took on a life of its own, sustained by funding from local and county governments.

To raise the consciousness of our volunteers about healthy wetlands, Mark and I taught them the diverse calls of different species of frogs so they could visit their wetlands after dark and identify which frogs were reproducing. If people heard the frogs calling and learned about the species, perhaps they'd care more deeply about wetlands.

Before I'd become involved with deformed frogs, I'd made tape recordings of frog calls at night and created an audio tape to give to the teachers I was training to identify wetland invertebrates. The calls were arranged on the tape in the seasonal sequence of breeding by the different species, from April to June, with help from staff at Cornell University's Library of Natural Sounds.

To capture the various species calls, I visited several wetlands on different nights with my tape-recording gear, sometimes after a long day at work.

One night I drove down a dirt road inside a wildlife area named Sunrise, located north of the Twin Cities, then parked and pushed my door open slowly, so I wouldn't scare the frogs. A black moistness swirled around me, and my ears reverberated with the intense high-pitched chirping of frogs: spring peepers, the harbingers of new life, too far gone in their annual mating ritual to sense my presence.

I stood up, feeling slightly dizzy in the night air, which was completely dominated by this primitive, amphibian passion. Early settlers had complained that they lost sleep because of frogs calling in spring. Now I understood why. I remembered a time I'd listened to peepers with my dad in New England, and, more recently, to the voices of Minnesotans, who sadly told me they no longer heard them chorusing in spring.

I booted up, shouldered my pack of taping equipment, and pushed through the trees and shrubs to the darkened wetland's edge. Mosquitoes whirred around my face. Cautiously I stepped down into the wetland, which, with its big hummocks of sedge, had very tricky wading, even when I had waded there during the day. Stumbling in carefully, I found my footing and pointed my shotgun mike across the wetland to the peeper-filled trees on the far side.

Why do people hear fewer spring peepers, I thought as my recorder spun for a good twenty minutes, the red signal lights bouncing up and down with the piercing calls. Is it mostly because we're losing wetlands to development and farming? Are misguided, turf-loving home owners and communities cleaning out the leafy, woody debris under trees that peepers need on the ground for cover during winter?

I have had little emotional attachment to frogs for most of my life. During college, I valued them highly for what they taught me about anatomy, embryology, and evolution, and now I saw frogs as important indicators to teach people about environmental health. But emotional connection? Frogs, let's face it, are cool and clammy to the touch, not soft and furry. You don't pet a frog. They leap away if you get too close, or dive under water in a pond, perhaps emerging farther away with only their black, beady eyes peering at you above the water's surface.

But one moonlit night that changed. I went to another wildlife reserve, a park northwest of the metropolitan area, spun in the code on the gate's padlock, and drove slowly along a sand and gravel road through the woods. The road curved up a hill into a restored prairie area, where I parked and walked quietly down to one of our most biologically diverse, high-quality wetlands, clearly lighted by the full moon. Beside me the prairie was bleached white in the luminous light; on the far sides of the glimmering water stood dark oak trees.

Close to the shoreline's emergent plant fringe, a pair of magnificent trumpeter swans, stunningly white in the moonlight, glided elegantly in unison over the calm water. Seeing me, they turned slowly away and floated effortlessly across the pond.

I stood in silence. Then I heard the low-pitched, rumbling call of leopard frogs start up in the near-shore vegetation. I was glad the wind had subsided; finally the frogs were audible enough so I could make a decent tape. Across the pond a symphony of spring peepers, chorus frogs, and tree frogs provided background. Perfect.

Something shifted in me that night. A deeper emotional connection settled in to stay, one that had been quietly developing during previous evenings out listening to frogs. I became mesmerized by their sounds.

In a dreamlike state, I pointed the mike to the water. This taping project was becoming less of a job and more of a spiritual experience. The frogs seemed to be calling me back across time, long before humans ever existed. For how many millions of years had male frogs called to females? For how many thousands of years since the glaciers retreated from Minnesota had frogs been chorusing in wetlands in spring?

After this experience, when the deformed frogs appeared in Minnesota, their plight quickly captured my heart and my work life. I saw the deformities as an atrocity inflicted on inspiring animals that are far more ancient than humans.

Another night I approached a wetland through a marshy field full of fireflies and stood quietly on the uneven shore. The frogs, sensing my vibrations as I'd approached in the dark, had stopped calling. While I waited for them to restart their chorus, I thought about the life I knew was in that pond: dragonflies, mayflies and caddis flies, *Daphnia,* fairy shrimp, snails, and beetles. I felt content just being there. Behind me lay an old farm field and a few shrubs and trees, and across the pond loomed an indistinguishable, black mass of woods where I once scavenged delicious morel mushrooms.

Then came the back-shivering trill of the American toad, full of spirit, seemingly heaven sent. The call lifted my soul. I was flooded with emotions and a kind of primitive, childlike kinship with nature that reached back to my past, long before I'd become so serious about my life. The soaring songs of white-throated sparrows I had heard in the Vermont woods entered my subconscious, as did the haunting calls of whippoorwills we'd hear at dusk while camping and my early enchantment with little black tadpoles (probably toads) that swam close to the shore of a pond near my home where we'd skate in winter.

Then another toad started its soaring trill, and another. The sustained, single notes of individuals overlapped each other and joined together magnificently in the moist, mud-and-water-scented night air. The chorus of toads hovered protectively over that beautiful, biologically diverse wetland.

My spirit reverberated with their calls. Then I understood why, after my father's death years earlier and before I started to work at the MPCA, I had tramped out to that same wetland to find solace. I sat there quietly for a long time thinking about how he'd helped me navigate some rough waters in my life and wished he could always be there. The spirit-filled calls of the toads calmed me; they transported me back to places Dad had enjoyed with our family in Vermont—to beaver ponds, lakes, and streams. I left feeling more at peace but still grieving his loss.

CHAPTER 5

THE PERIL WIDENS

With renewed hope, Mark and I made plans for surveying frogs and sampling ponds. In 1995 we had logged hundreds of misshapen frogs in a few different locations in the state. But what if Granite Falls repeated itself and no deformed frogs appeared in 1996? If that happened we could be accused of wasting taxpayers' money, of raising a false alarm.

Doubts hung over our plans. Getting out to frog ponds in time was uncertain at best. We were both absorbed by an onslaught of queries about the frogs; we battled to preserve our other projects; we struggled against the tedium of the state's agonizingly slow process for even small technical contracts. I pushed my mental accelerator hard, sometimes literally. Driving late one icy morning, I raced to a government building for a day-long training session so I could qualify for the new voice-mail system. I fumed: couldn't they just hand us a manual and have a contact person available to answer any questions?

Rarely could I stop thinking about the frogs, which, thankfully, the media kept in the public eye. One news article ran a photo of Mark and me in the MPCA biology lab; looking deadly serious, we're pictured holding jars of pickled, deformed frogs, like a Grant Wood portrait of a stark farm couple.

Late in March, Verlyn and I drove to Milwaukee, where I gave a talk about the frogs at a meeting on the amphibian decline held at the Milwaukee

Public Museum. On the way back across Wisconsin, we talked more openly about where we'd like to live. Even though we'd not yet committed to marriage, we had had started looking at houses in areas of St. Paul that would be close to my workplace.

At the end of April I flew to Boulder, Colorado, to speak at the EPA's Biological Monitoring meeting. I stayed at my son's apartment. Steve, a hydrogeologist, had just started his job at a consulting firm, where he would model pollutants released by mining operations into streams. When he'd interviewed for the position, he'd asked his new boss, "You're not someone who believes that heavy concentrations of metals like arsenic are there only *naturally,* are you?"

While I was at Steve's, Verlyn called from Rapid City where he was helping his elderly mother. "Are you ready for this?" he asked, pausing for effect. "Will you marry me?" For some strange reason, I laughed uncontrollably, mostly out of happiness. We were soon engaged and agreed to hold the wedding in the fall, knowing that I'd be away doing fieldwork in wetlands for most of the summer.

My fears about what might happen the coming summer ran deep. Could I handle everything: our intensive project to train volunteers, several presentations at meetings both local and out of state, and speaking to faith communities on weekends? After one talk I gave to a coalition of churches and environmental organizations that spring, a Catholic sister, who held deep convictions about protecting the environment, sent me a thank-you card. Inside it said: "The candle consumes itself as it serves others by shining." What did this mean? Was I consuming myself while I served the public interest? Shining? I was grateful when Sister Gladys called me periodically at home and encouraged me to keep going.

As summer approached, work accelerated even more and I struggled to keep up, spending time on weekends and staying late into the evening, sometimes after the security guard turned off all the floor lights, not knowing I was still at my desk. I had to find the appropriate switches in the closet of circuit breakers. I brought in a flashlight.

"Mark," I said one day. "How is it that I have eight talks to give on the frogs and wetlands in less than two weeks?" I had commitments to talk at a nature center, to the legislature, on an EPA conference call, at the Society of Wetland Scientists, at our division retreat, to TV's Nick News at the Ney farm, to staff at the state Health Department, and to a group that Verlyn belonged to called People of Faith Peacemakers.

Mark shook his head and asked "How much do you want to give up for this job?" I couldn't answer him. I had to admit that I drove a lot of what I did, but I couldn't see how to lighten the responsibilities. Cutting back on either the frog investigation or our wetlands work was out of the question. But how, given the intense demands of the frog investigation, could I continue my main work, the development of an invertebrate biological rating system for wetlands quality? Would I have to choose between that long-term goal and the immediacy of the deformed frogs? But how could I when both issues, both projects, were so important?

In early June I drove at high speed to the Ney Pond and donned my waders to be interviewed for PBS's *Newton's Apple* science program with the students and their teacher. Then I rushed back to MPCA, changed clothes, and made it to the state legislature just in time to testify on my newest grant proposal: to search for evidence of endocrine-disrupting chemicals affecting fish in streams and frogs at our study ponds. I told the legislators that these chemicals had feminized male fish in the Mississippi River, deformed alligator penises in Florida, and were suspected of increasing breast cancer and reducing human sperm counts. They could be harming the frogs, I said.

On a Saturday in early June, we taught several teams of citizens how to sample wetlands by wading into ponds close to the school where our first training session took place. In the school's biology lab, we instructed the volunteers on how to identify the different kinds of invertebrates and then use their findings to evaluate a wetland's health. These enthusiastic nonscientists had a lot to learn in only one day! I insisted we have biology teachers lead the teams, that we must pay them a small stipend. Several of these teachers had taken a workshop I had given a few years earlier, so they had some background and were already interested in biological monitoring. I knew they'd be able to help their team members identify the invertebrates they'd collected from local wetlands.

One volunteer marveled at the diversity of invertebrates he'd netted from a messy-looking wet area that had stumps and dead trees. "Who would have known?" he asked, astonished by what he saw in his sample tray. "What are those, are they little clams?" he asked, pointing to the tiny, pinkish-white creatures scurrying around in the tray.

"They look like clams, but actually they're crustaceans called 'clam shrimp.' They have leglike appendages, which are pretty much hidden inside the shells," I said.

I smiled when another volunteer told me how much fun her team had had in May when they went out at night and listened to the frogs at their study sites. "It was awesome," she said. "At first it was hard standing still for several minutes as we waited for the frogs to start calling. But then we loved it. And we could really tell the species apart!"

The Sunday after the all-day training session, I drove to a national wetlands meeting in Kansas. As I rolled through endless fields in Iowa, I reflected on the message in Sister Gladys's card. Would my flame burn out before my work was finished? At the conference, the interest shown in the biological scoring systems Mark and I'd created for evaluating wetlands energized me. At least EPA staff and other wetlands scientists were excited about our work, if our bosses weren't.

In mid-June, the prime time for MPCA biologists to be outdoors doing fieldwork in different areas of the state, the agency essentially incarcerated the entire staff of several hundred for a day-long session of customer service training. Certainly dealing politely and effectively with the public and regulated parties was a good idea. But would the environment now be second class, taking a back seat to customer satisfaction? Who was the MPCA going to serve in the future?

Customer service, we were told, was a "blend of practices and attitudes that maximize responsiveness to customers by understanding their needs and expectations." Key customer groups were "end users, those who directly use our products and services," and "beneficiaries, those who benefit from our services." The agency, it appeared, was shifting to a business model that aimed to please the people who were affected by the MPCA's activities and regulations: industries, municipalities, and citizens.

The goals we held for biological monitoring of the health of wetlands and streams seemed completely divorced from this new customer focus—unless the biological environment could be construed as a "beneficiary" of our service.

By July of 1996 young frogs had started hopping out of wetlands to find bugs to eat under low shoreline plants. We drove to the Ney Pond to conduct the year's first full survey of newly metamorphosed frogs. I walked along the vegetated shore and open areas near the pond and netted frogs, which I placed in the clean pillow case that hung like a sack from my belt, a handy method I had learned from Bob McKinnell. I'd been buying used pillow cases of various patterns and colors at thrift shops and laundering them without detergent, to protect the skin of the frogs we collected. Occasionally an unhappy frog croaked inside my sack, which swung from my waist as I moved.

We gathered back at our vehicle in the shade, sat on coolers, and carefully examined each frog before we released them gently into the low vegetation near the pond. One of us observed, the other recorded. Over and over I gave the frog's length, then said "normal—normal—normal." To my astonishment, only one out of the 124 frogs we had collected that July day looked deformed. It had a layer of skin over one eye, but otherwise all its other parts, legs included, appeared fine: nothing compared to 1995, when a third or more of hundreds of frogs we had collected at the Ney Pond in late summer and early fall had obvious, gross abnormalities.

Was this to be Granite Falls all over again? By this time students, teachers, the media, other biologists, and the general public were paying close attention to what we did. Everyone had been alarmed about the deformed frogs in 1995. They expected us to investigate. For me to remotely *want* to see deformed frogs at the Ney Pond seemed twisted. Was I like a medical researcher who worked on a particular human birth defect, and wanted to have more examples of it to study, more defective babies born? More crippled frogs to appear?

Back at the office, I listened to my voice-mail messages. One came up marked "urgent," a woman saying, "My frogs are all bad. Please call me"; another from a teacher in northwestern Minnesota. He'd found several frogs missing parts of limbs or entire legs. One caller reported seeing lots of deformed frogs in north-central Minnesota; another found at least 5 percent deformed frogs at his lake property. One woman said her daughter had a frog with a missing eye, found near Lake Minnetonka west of the Twin Cities. A caller spoke urgently of frogs with missing legs and "the bladder outside its body." As the summer of 1996 progressed, Mark and I were inundated with reports of deformed frogs.

A woman living in the west-central part of the state called. Her kids had caught frogs and put them in a pail to show her. "Our frogs have little stumps for legs," she said. "There's no way these were injuries from mowers. For one thing, we didn't use any machines this summer, and besides, I saw no blood or cuts on the legs. This is scary to me. What is going on?" she asked.

I asked about the landscape around the wetland. She described the runoff that floods their land from a farmed area nearby. "In spring there's a river of water that comes off the soybean fields and flows to the sloughs," she said. "But also, the farmer there puts septic stuff from the pig farm on his fields."

Sadly, my bizarre, fleeting wish for deformed frogs to reappear had come horribly true. Once again, frogs with deformities were emerging—but this time all over the state. I created an intake form so we could record information from anyone who called us: the date, the location, the number of frogs, the kinds of deformities they'd seen, their contact information. Had the person looked at frogs in past years? Had they seen any deformed frogs before this summer? Would they or the landowner be willing to let us come and survey their frogs? Was it only one species?

From here on, I could think of nothing but these pathetic frogs. How could they move about and eat with such crippled legs? How could they escape predators like herons, raccoons, and garter snakes ready to dine on frogs?

Each day as I pulled into the staff parking lot back of the MPCA building, I'd swell with trepidation over the pathos, fear, and horror my voice mail might reveal in messages from people who were finding deformed frogs. I almost welcomed the slow ride in the cranky old elevator to the fourth floor, a ride that normally drove me crazy. What would today bring?

I'd sit down at my desk, grab a stack of recycled sheets of paper, write the day's date on the top, then lift my phone to listen to dozens of voice messages. Many new locations of deformed frogs were coming in; *too many.* When I was through, I highlighted priority calls in pink, names of citizens who'd reported extraordinary numbers of deformed frogs or other key people—like scientists—to call back as soon as possible.

One day I grabbed my mail and opened a hand-addressed envelope containing an article about a two-headed girl, conjoined twins born in 1990 in a rural Minnesota town west of the Twin Cities. They were pictured happily riding a bike, their smiling father standing close by. Much later I learned these girls, now young women, would be able to drive a car and attend college. Conjoined twins probably derive from an error in the separation of the earliest cleavages in the developing fertilized egg, although other theories exist. I doubted this had any connection with what was causing malformations in frogs. Still, I stared at the picture, marveling at their apparent happiness . . . and their dad's.

As citizens poured out their stories, concerns, and questions, I could not numb myself to the repeated news of abnormal frogs. There was a kind of fascination over each novel description of deformities. Many people reported variations on the missing and partial limbs that we'd so frequently seen; some people saw frogs with extra legs or missing one eye. Incredibly, two individuals reported a frog missing *both* eyes. One called about a frog with three legs on one side for a total of four rear legs. We had other reports: a frog with a leg coming out of its neck; a frog with an extra leg in front; a frog that had rear legs that wouldn't work so it couldn't hop. One person saw a one-legged frog hopping around in a circle. One frog had what a caller described as a "cone head"; another had gills and a tail but full legs and moved very slowly.

It was not easy for some people to talk with a government worker they didn't even know, to identify where they lived. I could hear caution in their voices. Underpinning people's anxieties lay the fear that something insidious might harm their families; that something bad was happening in the environment where they lived. Understandably, they wanted answers—what's causing this, what should we do about it—but talking with these citizens was difficult because we had no easy answers and couldn't minimize their concerns.

In 1995 the deformed frogs had provided a stunning news story about kids finding weird frogs at two ponds near Henderson and Litchfield. But now, in 1996, the problem was expanding like a scary epidemic of some new and mysterious disease spreading across Minnesota.

Mark and I hurriedly planned to visit some of the reported sites, if for nothing else to establish locations for others' research the following year. Along with many biologists, we struggled to make sense of it all and to develop a decent study design. We should include ponds in different areas of Minnesota, not just focus on the Ney Pond.

Could we establish and investigate paired sets of wetlands, one with normal frogs to serve as a reference site, the other with deformed frogs? Establish pairs of sites in a few regions of the state? We hoped that differences would emerge between the "normal" vs. "affected" ponds, the latter those known to have deformed frogs. A staff person with statistical expertise affirmed that our paired-site study design was valid. We'd sample and chemically analyze the pond water and sediments, survey the frogs from each site. We'd send some frogs out for further analysis to look for diseases, parasites, and, if funding was sufficient, contaminants in their tissues.

In July, we surveyed the frogs at a pond located in an attractive lakeside area in central Minnesota near the town of Darwin. As we rolled off the highway down a gravel road that ran parallel to the shore of a large lake, we approached the tiny, low-lying pond edged with woods on its upslope side. Harmless looking, the pond sat nestled close to houses and cabins that faced the lake. A pleasant area to live or vacation.

We walked casually around the sloping lawn near the pond's edge, swept our handheld nets to capture frogs, then grabbed the net with the frogs leaping inside and guided them into the pillowcases that hung from our belts.

The collecting was easy; the viewing was not. I sat down and pulled one tiny frog after another from the mass of frogs in my pillowcase and was sickened by what I saw: frog after frog with a missing a leg, a partial limb, or one abnormally bent and twisted. Few were normal. Our crew sat working in silence, speaking only to record the deformities, a grim task. The majority of the frogs were deformed, 75 out of 112. Two-thirds. Far beyond my wildest thoughts about what might happen that summer.

Now I lived in a real nightmare. Long past my short-lived wish to have deformed frogs reappear that summer, my fears deepened as I viewed too many misshapen bodies squirming in my hands. Sometimes I just had to look away, inhale the breeze of lake-scented air, and wonder what all this meant.

Most of these physically impaired frogs couldn't possibly survive to the next season. When one appeared normal, I felt grateful. At least that frog might successfully hop into the lake for the winter.

We headed to northwestern Minnesota to survey a site reported to us by a citizen who'd been out collecting frogs to use as bait for a local catfish tournament. The wetland was located near a nineteenth-century steamboat stop on a major tributary to the Red River of the North, which flows from Minnesota north into Canada. We collected 11 deformed frogs out of 117, or roughly 10 percent deformed. Many lacked a hind leg or part of it. As we tromped around, netting frogs, a crop duster airplane flew over low, spraying pesticides on a nearby field of sugar beets, planted right down to the edge of the pond. Uneasy about the spraying, we retreated temporarily to our vehicle.

At another pond in the northwest, we found 37 deformed frogs in the total of 239 we had netted, at least 15 percent of this population was deformed. By now we were calling such ponds "hotspots." The only chemical the land owner reported using was the herbicide Roundup, but the immediate area surrounding the pond had crops, the owner said, and pesticide use there was heavy.

At this site, the owner's children were eager to help us. The kids said they hadn't seen deformed frogs before, and I believed them. They'd remember if they'd seen such weird-looking frogs in the past.

We began our quest to find ponds where the frogs were normal so we could pair them up with a hotspot site located in the same area. In August we trekked north to a rural pond next to the home of a Natural Resources Department employee. Her children had been collecting frogs in previous years, and they reported that none of them had looked abnormal. "Nice frogs," I wrote in the field book as we recorded 155 normal frogs.

A key issue was the background rate of natural deformities. I asked zoologist Dave Hoppe what he thought. "I've seen maybe one or two deformed frogs in ten thousand," he said, reflecting on his many years surveying frog populations in Minnesota. Later, Dave's analysis of museum collections confirmed what he knew from his extensive field research: seeing so many deformed frogs in the mid-1990s was not normal. Biologists would debate what percentage of frogs with abnormalities might serve as a reference point for a typical "background" rate. What observed increase in frequency would be deemed unusual? Some proposed 1 percent (one in one hundred frogs) as a minimum background; others, 3 percent. We tried to collect at least one hundred newly metamorphosed frogs in each survey, so we could detect these low percentages.

The concept of background rate does not necessarily mean that developmental malformations are natural and normal, although genes are known to have a background rate of mutation. In humans, approximately three to four babies are born per one hundred live births with an abnormality. But in less developed countries, where pregnant women are more vulnerable to diseases and vitamin deficiencies, the rates of birth defects are higher.

Early in September, we returned to survey the frogs at the Ney Pond, and this time things were different: the deformed frogs were back. A one-legged frog flopped over and exposed its white belly as it tried to escape our nets. I watched it struggle on the mudflat near the water. Like a turtle, it lay helpless on its back, its single leg flailing, its white belly an easy target for predatory birds.

Another frog lacking a leg swam in circles in the turbid, brownish water, like a boat being rowed with only one oar. Usually escaping frogs will jump into the water and dive down immediately and hide underwater. Not this one.

Of the 104 frogs we found at the Ney Pond, 8 were deformed, roughly 8 percent. Yet two months earlier in July, we had found only one mildly deformed frog among the more than 100 frogs we had collected. Something clearly had changed there, but what? Was the development of deformed frogs delayed by something in the water while the normal larvae were able to grow faster? Did the numbers of deformed frogs reflect variation in the quality of eggs from different females? Might some egg masses be laid at critical times when parasites or pollutants were in the water?

Compared with deformities we saw at Ney in 1995, this time very few had extra or branched limbs. Extra-legged frogs had become an icon for the deformed frogs, like the frog I photographed lying on its back with two slender extra legs sticking out of the lower abdomen between the two normal legs. But by now we knew that such abnormalities were infrequent. Far more frogs had missing or partial limbs than duplicated ones.

By the end of September of 1996 we had received reports of malformed frogs from an astounding 172 different locations in Minnesota, in 55 of the state's 87 counties. As each report came in, I marked pink dots with a highlighter on a state map, which began to look as if an epidemic of deformities had hit Minnesota.

Our surveys verified twenty-two sites reported from citizens that year. Not one report was false. Every site had deformed frogs. We computerized a map of the state with outlines of the eighty-seven counties and black dots to show each location of deformed frogs that had been reported. The map of Minnesota looked ominous.

Most of the deformed frog sites were located within a wide swath from southeastern and central Minnesota ranging toward the northwest. This region is populated with wetlands and hard-water lakes, prime frog habitat, an area that historically had deciduous hardwood forest. In EPA lingo, it's the North Central Hardwood Forest Ecoregion. Mark and I knew this large area of the state well because we had sampled many wetlands in that region to develop our biological scoring systems.

Only a few deformed frog sites were reported to us from northeastern Minnesota, an area where coniferous forests are studded with soft-water lakes edged with massive granitic boulders. Few reports came in from the predominantly agricultural southwestern corner of the state, a region where 90 percent of the wetlands had been drained to convert wet areas into cropland. To create more tillable land, drainage systems ran water from smaller wetlands, the type preferred by frogs, directly into rivers or into larger shallow lakes, making them deep enough to support fish but few frogs.

During the summer of 1996, EPA staff started organizing a meeting with US and Canadian scientists to discuss the frog deformities, to be held in late September in northern Minnesota at their lab in Duluth, located next to Lake Superior. I looked forward to a gathering of scientists where all ideas could come into play. By then the deformed frogs intrigued and mystified many researchers. Perhaps they could work together to begin to solve this mystery.

Late in September I drove my Toyota north from St. Paul to speak at the EPA's meeting of scientists assembled to discuss the deformed frog phenomenon. As the Duluth harbor and the oceanlike expanse of Lake Superior came into view, I wanted to keep on driving past the EPA headquarters and head northward along the great lake's awe-inspiring shore. If only I could walk across a cobble-covered beach and pick up a couple of hand-sized oval stones, the type my boys used to call "smoothie stones." I yearned to hear the magical sound these wave-smoothed rocks make when they clink together in retreating surf—like crystalline glass. Might this erase some of the summer's horror of deformed frogs?

Instead, wired up and expectant, I drove to a motel in Duluth for the night and rehearsed the talk I'd give the next day to EPA's assembled experts.

Would the scientists take this problem as seriously as I did? Would they suggest ways to find some answers?

In the morning, zoologist Dave Hoppe began the meeting. He spoke soberly. Grim-faced, Dave told the audience that he'd observed at least two hundred malformed individuals among *six* different species of frogs in the past two summers. "Before this," he said, "At most I've seen a couple of malformed frogs in thousands I've collected over the years. *This is not normal.*"

An acute observer, Dave could spot fingernail-size, newly metamorphosed toadlets, which to me were merely black specks in the grass. At the Ney Pond that summer, Dave suddenly stooped over to catch a tiny toadlet, then gently pulled its leg. "This one has a truncated limb," he said, as he and I scrutinized the miniature frog up close. He took some of them back to his lab to examine more carefully under the microscope.

At the meeting, Dave Hoppe proposed his theory that species of frogs whose tadpoles spend longer periods of time in pond water might develop more deformities than frogs whose larvae mature faster. The longer the development time, the more exposure to water-born pollutants or other agents, he reasoned. "I've seen this at one hotspot site this summer," he said. Dave had observed more deformities in mink frogs, whose tadpoles remain in the water almost a year before metamorphosing, than in tree frogs and toads, which need only a couple of months to develop.

Next, Bob Dubois, a biologist from Wisconsin's Department of Natural Resources, stood and reported finding deformities very similar to what we had seen in Minnesota, but in *seven* species of frogs. People looked up: having deformities in that many different species was very significant. He saw no "obvious common denominator" among the various places where they'd collected deformed frogs in Wisconsin: some were found in relatively undisturbed areas, others in urban and agricultural landscapes. "I've talked with experienced field biologists," he said, "and all of them feel this is a new occurrence in our state."

Then Martin Ouellett, a lively researcher from Canada who practically bounced with youthful energy as he talked, stood up to tell us they'd found deformities in an astounding *eight* different species of frogs in the St. Lawrence River Valley in Ontario, beginning in 1992. Even those of us who had, by then, seen lots of deformed frogs were sickened by some of Martin's slides, especially one hapless frog with an eye bulging out from the side of its body. We stared at the horrific image on the screen. Someone near me gasped, "Oh, my God."

That summer Bob McKinnell's technician at the University of Minnesota had tried unsuccessfully to feed bits of liver to a small deformed frog in Bob's lab, but the frog wouldn't eat. The technician pried open its mouth and shined a light to look inside. An eye protruded into the throat. How can eyes be so terribly misplaced?

Martin had seen the whole range of defects in Ontario frogs: amputations, extra limbs or digits, rigid legs, back-folded (or anteverted) limbs, atrophied muscles, skin webbing (or cutaneous fusion), bony projections, missing eyes, deformed jaws—even a blue frog.

The Canadians were way ahead of us in their frog investigation, and I couldn't help but feel a bit envious. Already they were analyzing frog tissues for metals and other chemicals, diseases, parasites, and cellular DNA content. They'd surveyed farmers about their pesticide use and were analyzing pond water for chemicals. In his lively French accent, Ouellet summarized his results: he found more deformed frogs in ponds surrounded by agricultural fields than in the nonagricultural reference ponds. He pointed to pesticides as a probable cause of the deformities.

How had the Canadians whipped into action on this problem so quickly and initiated so much sophisticated analysis? True, they'd had their first really significant outbreak of deformed frogs three years before Minnesota's. But were they also somehow freer of the kind of political and governmental roadblocks we kept encountering? Was their government more responsive to a new environmental problem than ours?

In the 1990s the biennial budgets constructed by MPCA managers left little leeway for tackling unexpected issues like malformed frogs or new types of toxic chemicals showing up in lakes and fish. Years later the agency would assign staff to work on emerging issues, such as endocrine-disrupting chemicals and other newly discovered problems. But it remains to be seen whether that work will be funded in the future and given the tools it needs to move forward and monitor and regulate the new types of pollutants.

I learned the most at the Duluth meeting from Ken Muneoka, a developmental biologist from Tulane University. Dr. Muneoka carefully described experiments that elucidated the embryonic mechanisms for limb development. "Rearranging some tissue in the early limb bud stage of developing tadpoles can trigger multiple limbs," he explained. Applying a chemical called retinoic acid to the limb bud after some tissue was removed would also cause extra legs to sprout, he went on, displaying slides of multilegged frogs. Parasites were not the only explanation for branched or multiple legs in frogs. Chemicals could trigger them too.

Alternatively, he showed that removing cells from the tadpole's early limb bud could cause shortened or truncated limbs to form. Applying particular chemicals to the budding leg could inhibit its outgrowth. There we had it—branched and extra legs as well as shortened legs—all created in the lab by chemicals. I was fascinated that such basic knowledge about development was being revealed at the cellular and gene level.

Dr. Muneoka concluded: "Certain chemicals which are able to disrupt the early signaling of genes could cause very different kinds of malformations, depending on *where* they contact the early tissues." But birth defects can also develop differently when an embryo is exposed to the *same* chemical disrupter at different times, he said. Timing of exposures as well as its tissue location are both critical determinants of the kinds of defects that develop.

When he told us that methoprene, an insecticide used to control mosquito larvae, could cause deformities in animals, he had my attention. Methoprene was widely used around the Twin Cities. I knew it negatively affected other wetlands invertebrates as well. Methoprene is structurally related to retinoic acid, the normal developmental signal that he'd been discussing.

Deb Swackhammer, a toxicologist and environmental chemist from the University of Minnesota, talked about deformities in Great Lakes fish-eating birds, mammals, and snapping turtles. Some of the deformities had been attributed to chemicals deposited in the lakes from the atmosphere, she said, others to industrial discharges into lake water. I knew a bit of this story, but still found it outrageous that rainfall landing on the almost unpolluted waters of Lake Superior could bring with it toxic contaminants that harmed the wildlife that lived there. Previous studies had shown that pesticides in rainfall are at their highest concentrations during the spring in Minnesota, the time when pesticide applications on crop lands are heaviest, the time when many frogs reproduce.

Joe Tietge and Gary Ankley, both EPA scientists, raised the possibility that the increase in ultraviolet light (UV), resulting from the expanding hole in Earth's protective atmospheric ozone layer, could cause deformities in frogs because ultraviolet light directly damages genes. Amphibians have some built-in protection against normal levels of UV radiation in their eggs, but this safety mechanism could be overridden by too much UV landing on eggs as they floated near the pond's surface, he said. The scientist pointed out that UV light can also cause damage by converting or photoactivating a harmless chemical into a toxic form, possibly one that causes deformities. A puzzling fact was the steady—not sharp—increases in UV radiation over many years, compared with the sudden eruption of deformed frogs during the early to mid-1990s.

Then a zoologist from New York, Stan Sessions, talked about his work on a parasite whose adult stage is a liver fluke in vertebrates. The immature stage can invade tadpoles after being released from snails, which serve as an intermediate host in the parasite's life cycle. If the parasite invaded the tadpole during early limb development, he said, it could trigger extra limbs. He'd found cysts of parasites in multilegged frogs from a pond in northern California. In the lab, he'd implanted resin beads into limb buds of tadpoles, and this had triggered extra legs. He theorized that the beads physically blocked the movement of the normal chemicals that controlled limb development. I wondered if the resin beads themselves might release a chemical. At the time, trematode parasites had not been shown to cause limb malformations directly by exposing tadpoles to immature stages of the parasite in lab studies.

The life cycles are complex. Eggs of adult liver flukes that grow to maturity in birds or mammals are shed in feces into ponds, hatch, and invade their first intermediate hosts, certain snails. Within the snail's tissue they develop further and later hatch and shed from the snails in the water. These stages then swim to find the second intermediate host, either certain fish or amphibians, penetrate their tissues, and encyst as a resting stage. If the fish or amphibians are eaten by birds or mammals, the encysted stages develop again into the adult liver fluke to complete the cycle.

Some biologists, myself included, were skeptical about this hypothesis. Why so many parasites now and not earlier? Others pointed out that human activities, such as excess fertilizers running into ponds, could have caused an increase in the parasite's snail hosts by promoting growth of the algae and plant material they feed upon. But such runoff had likely been going on for a few decades, long before any surges in malformed frogs occurred. (See epilogue for more recent parasite studies).

For people wanting a natural hypothesis for frog deformities, this was it. Sessions told a reporter that he agreed that environmental chemicals could also play a role in the deformities. But from 1996 on he became an ardent, if not aggressive, proponent of the parasite hypothesis as the main cause of deformities. Later he wrote me that we don't really need any other cause than parasites to explain them.

When I got up to talk, I displayed a transparency of our map of Minnesota to show the locations of the 172 deformed frog sites reported to us or actually surveyed by our field crew, and an example of the intake form we'd devised to take down specific information from citizens who called us about deformed frogs. When I did this, I noticed a couple of scientists up front leaning together and whispering as I talked.

During the follow-up discussion period, the two scientists said that data from citizens couldn't be considered valid and shouldn't be used. "This is not the way to study the scope of this problem," one said. "Instead, a randomized survey of ponds should be undertaken."

Hearing this, I felt perturbed, not by the idea of using a randomized study design, but at the veiled insinuation that the citizens who called us were not telling the truth. I knew that our field collections verified their reliability as reporters about what they'd seen near their homes. The implication that our work wasn't "scientific" enough, when all we could do at that point was scope out the problem in the field, put me on edge.

Finally, feeling a bit defensive, I stood up and said to the assembled group that if we had not had all those reports from citizens we might not even be having this meeting. "Without their reports, we certainly wouldn't have had a clue about how widespread the deformed frogs actually are in Minnesota," I said. We had surveyed twenty-two of the sites phoned in to us by citizens that summer and not one of those reports turned out to be false, I said. Minnesota is a huge state. Given our small staff, we couldn't possibly have discovered on our own so many locations that had deformed frogs.

Later, when the EPA posted a summary of the Duluth meeting on its website, I saw a statement about reports from citizens: "Data collected by untrained individuals may not meet the objectives of future studies. Furthermore, it is difficult or impossible to enforce data quality assurance."

Of course I understood that a landscape ecologist's methods and random surveys of ponds would be needed to truly evaluate the *extent* of the deformed frog problem, to predict how many ponds might have deformed frogs, and ultimately to assess the overall impact of the deformities on frog populations. But I couldn't see how knowing that would help us understand the *cause.* Deformed frogs were widespread. We already knew this from our surveys, from the work in Ontario, and from reports we had received from other states. To get at causes, especially with limited resources, wouldn't it be better first to research carefully some ponds that we already knew were producing deformed frogs? How do those wetlands differ from sites where the frogs are normal?

Various approaches to explore causes of frog deformities were discussed at the wrap-up of the two-day meeting in Duluth, like rearing frog eggs in cages at ponds that had generated deformed frogs, while at the same time rearing eggs in clean water in lab aquaria. Would that pond water trigger deformities, while clean water did not? This would suggest the causative agent was in the pond water. Effects of UV radiation on developing frog eggs needed study, and histological examinations should be undertaken to confirm that shortened limbs were not the result of physical injury. Also a spectrum of frogs, both normal and abnormal, should be examined for evidence of diseases and parasitic infections.

I left Duluth feeling energized by all the ideas that were discussed. I joined up with Kathy Converse, an enthusiastic biologist who worked for the National Wildlife Health Center Lab in Madison, Wisconsin. Together we drove fast across northern Minnesota to connect with an MPCA field crew at a pond shown earlier that summer to have a high percentage of deformities in frogs; we wanted to get there so we could survey the frogs before dark. Kathy's goal was to take some live deformed specimens back with her to the Madison lab, where researchers would examine them for diseases, parasites, physical trauma, and possibly chemical contaminants. We talked openly about what could be done to grapple with the frog problem. As we drove, Kathy and I shared stories about our families. She was helping her daughter be more organized about school. I told her about Verlyn and the plans for our wedding in November.

I was glad the National Wildlife Health Center, a research facility under the US Geological Survey and located in Madison, Wisconsin, was planning to pitch in. Their scientists would play a key role in analyzing malformed frogs from Minnesota and other regions of the country.

The center was created in 1975 to analyze outbreaks of diseases in wildlife and to educate the public. Its labs are staffed with experts in pathology, parasitology, microbiology, biochemistry, and wildlife ecology. They're equipped to analyze diseases rapidly, particularly in migratory birds, endangered species, and other wildlife living on federal lands across the country. They work on outbreaks of diseases such as the West Nile virus, Lyme disease, and chronic wasting disease that could impact humans and harm domestic animals. The center has researched deaths of thousands of ducks and millions of other birds, as well as die-offs in amphibians and tadpoles.

When we met up with the crew, it was gray, cold, and miserable. Collecting was difficult, but we managed to net fifty-six frogs that chilly day, fifteen of them deformed, over 26 percent. That site remained a definite "hotspot."

On Friday after more collecting, Kathy headed back to Wisconsin, her live frogs carefully ensconced in small plastic aquaria. I climbed in with our field crew for the drive south to St. Paul. Late that evening, we arrived at the MPCA's darkened warehouse and stashed our field gear on tall metal shelves labeled FROG WORK. I headed over to the MPCA building with several live deformed frogs that I wanted to anesthetize and then preserve. Normally, I would have done this in the field, but we didn't have the drug along. Earlier I had vowed never again to subject frogs to the torture of dying in the formalin preservative without first putting them to sleep. Exhausted, I opened drawers and cabinets in a fruitless search for the anesthetic. Finally giving up, I decided to wait until Monday to preserve them. I left them in a covered aquarium in the lab, not knowing these frogs would have an altogether different fate.

As soon as I got home after nine o'clock, the phone rang. "Are you okay?" Verlyn asked, with a worried sound in his voice. He knew I'd been up north doing fieldwork that day and that the weather had taken a nasty turn. "Yes, but right now all I can think of is food and sleep," I said. He reminded me our book club was meeting Sunday evening, the last thing in my mind at that point.

CHAPTER 6

SCIENCE IN THE PUBLIC EYE

The Monday following the EPA's meeting in late September 1996, we conducted a third survey of the frogs at the Ney Pond. This time a troubling 32 out of 70, or *47 percent* of the frogs were deformed, much worse than earlier in September, when we logged 8 percent abnormal, or in July when only one frog out of 124 we collected had a small defect. With each survey the percentage of deformed frogs was increasing. How would people react if such a horror were happening to newborn humans?

Forty-seven percent! What was causing this, and what could be done to stop it?

Back late from the field trip, I decided to go up to my desk and quickly check for messages before heading home. Mark was still working at his computer as I walked into our work area.

"The *Washington Post* has an article today about deformed frogs and the Duluth meeting," Mark said. "You'd better look at it. What's more, CBS News showed up here today in response to it."

I pulled up my e-mail and saw that an article, titled "In Minnesota Lakes, An Alarming Mystery," had run on the front page of the *Washington Post* (Souder 1996). The article summarized the students' 1995 discovery of deformed frogs and reviewed the previous week's deliberations among researchers at EPA's lab in Duluth. Scientists had many ideas but no answers, and the possible causes of the deformities were "almost limitless." Local reporters jumped in (Lien 1996).

"Your phone has been ringing off the hook," said Mark. Sure enough, my voice mail was jammed with requests for interviews with local and national TV networks, newspapers, and radio stations.

Minnesota's frogs had moved onto the national stage.

Suddenly, like it or not, we were going to be thrust into the media spotlight, and I was not happy about the prospect of doing science in the public eye. I headed home, feeling overwhelmed by the avalanche that was about to cascade over our heads. I could see my life spinning even further out of control. How was I supposed to catch up on commitments like the overdue reports to the EPA we'd had to put on hold while out surveying the frogs and wetlands all summer long? How could I sustain my focus on the frogs when the media was about to suck away our time and energy to do that work?

During those first days in October, the exhilaration of all the media attention swept me unexpectedly into its seductive embrace, temporarily overriding my resistance. I had several deformed frogs still alive in the lab, frogs I had intended to preserve properly for analysis by our research partners. Instead, I offered them up to the visiting news crews, thinking at the time that this exposure might ultimately help our cause and that of the frogs.

When calls came up from the MPCA receptionist to alert me that a scheduled film crew had arrived downstairs I swelled with gratitude, and—I admit—some self-importance. Before this, I'd dreaded giving my work time to the media. But now the outside world was about to acknowledge that Mark and I were involved in a significant environmental issue, that deformed frogs merited attention. I knew many agency staff took the frog deformities very seriously; I just wasn't convinced our management did. We continued to feel marginalized, out there working on fringe issues like wetlands and frogs, not in the agency's mainstream.

I dashed down the windowless stairwell to greet the film crew and escort them to the biology lab in the basement, lifted the malformed frogs out of their aquarium, and placed them on a cart covered with a rubberized mesh material. Men shouldering large video cameras hovered over the tiny frogs to capture their feeble attempts at escape, their off-balance leaps that landed them belly up and helpless looking because of a faulty leg. Complicit in the whole process of exposing these frogs, I even nudged one to make it leap, knowing it would flop over for the camera.

The media has been described as a monster that must be fed, and for a few days I fed it with the live frogs that I had tried unsuccessfully to preserve just before the media arrived on our doorstep.

But then I had second thoughts. Displaying these handicapped frogs, even for their possible ultimate benefit, didn't seem right. It felt tawdry. Earlier, Mark and I and other biologists had committed to sacrifice only the frogs we absolutely needed for research on the causes of the deformities. All others we returned to their ponds. We had agreed not to take frogs—nor allow their taking—for frivolous purposes (like TV shows, I now felt). And that summer I had rebelled against a request for us to collect some deformed frogs to display at the MPCA booth at the Minnesota State Fair held in late August. This was an inappropriate use of these misshapen animals, I'd said. It smacked of the historic though long-banned freak shows staged at the fair.

Yet there I was, offering up our frogs and succumbing to the siren appeal of TV film crews and newscasters.

In my few spare moments Verlyn and I planned our wedding for early November, a simple one (we hoped) with an old-fashioned church basement reception. But the onslaught of media attention at work occupied most of my attention. How we could begin to merge our belongings into one place? Make plans?

After the media blitz, the live frogs became caught in a power struggle with the MPCA's Public Information Office (PIO), which wanted to keep them in their offices "for educational purposes." I suggested an alternative solution: that PIO staff film the deformed frogs in the lab, as TV crews had already done, so I could go ahead and preserve them for research. Instead, Mark and I were summoned to the information office on the first floor for a meeting titled "Frogs, Education, and Ethics." The agenda: to discuss the "transfer of frogs."

"I've talked about those frogs you have in the lab with the commissioner," the director of PIO said, looking at us sharply. "He agrees that our office should take over the frogs and keep them here."

I looked at Mark, who rolled his eyes. We were sure that the intent was to use the frogs to lobby for the agency's upcoming budget request to the governor and legislature, not for "education." I saw this as a cynical ploy to promote the MPCA's other work when I was quite sure the agency's leaders had no intention to continue investigating the frogs in 1997.

Finally allowed to speak, I said with clenched teeth: "The biologists working on the frogs have made a commitment to take frogs from ponds only for research. What's more, the agency has no guidelines for keeping live animals."

At the University of Minnesota, I said, Bob McKinnell, who used frogs for his cancer research, had to pass scrutiny from the university's animal care committee for everything he did with them.

To this the information office's director replied tersely, "I've had experience working with live animal displays with the Department of Natural Resources," she said, shifting in her chair.

"Did DNR ever display any sick or abnormal animals?" I asked.

She gave me a strange look, then replied slowly, "No they did not."

I pressed on: "If you were working for the Health Department, and were going to lobby for their budget—"

But I couldn't complete my sentence, because the director, her face red, was yelling over my words and saying, "I refuse to go down this path!"

I continued. "—would you take some deformed human babies to the Legislature as part of your funding pitch?" By now I feared her fury might make her explode.

"We *will* have those frogs," she seethed. Defeated, Mark and I walked out. I don't think I would have felt any different if I knew that the MPCA *had* included funding for the frog work in its new budget. But they had not.

I had made my point knowing I'd lost the battle. The deformed frogs were taken against our will from the lab to the information office and kept under the care of one of their staff people who had some background in herpetology. She was very friendly and reassuring about their care, but it bothered me deeply to see those frogs used as lobbying tools to make the MPCA look good. Granted, being hand fed was far more pleasant than a sleepy death in formalin, but to my mind the hapless frogs, kept like prisoners under bright lights in an overly warm and unsecured room at PIO, were being used inappropriately.

A year later, in 1997, the information office developed its Internet "Frog Cam," a video camera pointed at deformed frogs round the clock. People could log in from anywhere and see the frogs live on their computer screens. Did they do this to raise public consciousness about our troubled frogs or to bring attention to the agency? I was never comfortable with this display and felt no joy when the information office would proudly announce how many hits they'd had on the Frog Cam's website. The hypocrisy was too great.

The first reporting by local print and broadcast news journalists in 1994 and 1995 told the story of the deformed frogs without distortion. I'd welcomed local TV reporter Ken Speake's coverage that August day when I visited the Ney Pond for the first time in 1995. It even helped bring the MPCA's investigation into being, when the agency would rather have looked the other way. And during the spring of 1996, when we sought funding to enable us to pursue the frog problem, the media coverage kept the public's attention in a way that I never could, no matter how many talks I gave to diverse groups of people.

By the summer of 1996, however, TV networks and reporters were exerting pressure through the Public Information Office to accompany us while we collected frogs and took samples from ponds for chemical analysis. It became an impediment, not a help. Not only were there issues of landowner privacy, I feared the media would distract us from the work at hand. By then, we were developing our procedures for collecting frogs and pond samples, and often we needed to discuss what we were doing as we worked in the field. I refused such requests.

CBS News, then anchored by Dan Rather, led their coverage by asking, "A plague of frogs, a warning?" with a full screen showing a stump-legged frog (filmed in our lab) lying on its back. The word *MYSTERY* overlaid a pond with a blooming water lily. Then Bob McKinnell, garbed in a white lab coat, was shown holding up a deformed frog. He boomed emphatically: "When the frog that samples our environment becomes abnormal or dies, it is raising a red flag." Over a large picture of a live frog, its two legs shortened and flailing helplessly, the CBS narrator intoned, "Minnesotans worry. If contaminants are to blame, this biological mystery could end up the amphibian version of Love Canal." Love Canal was the infamously contaminated neighborhood near Niagara Falls where babies were born with birth defects because of toxic chemicals.

On NBC I was pictured holding a frog with a thick, stubby leg with an oddly branched foot structure at the end. "These frogs are sending us a very strong signal about the quality of the environment they're in," I said. Their camera panned around the rural landscape at the Ney Pond and the narrator stated solemnly: "It's happening in waters that don't look polluted—in Minnesota's pristine lakes, some surrounded by thick woods, others by rolling hills and lush farmlands. The beauty along Minnesota's waters is hiding a killer." I almost expected to hear the thumping theme from the movie *Jaws.*

The *National Enquirer* emphasized a pregnant Minnesota woman's concerns and a person who asked whether it was safe to buy a home near where deformed frogs were found (Ziegler 1996). At the time I had no satisfactory answers for the people who called us with legitimate fears like this, other than honestly saying that we had no knowledge whether or not humans could also be affected. As long as we didn't know what caused the epidemic of deformed frogs, we couldn't assure anyone that people would be free of harm.

Even the foreign press entered in: a Danish newspaper talked with us and later quoted one of the students who found deformed frogs at the Ney Pond as saying (in Danish), *"Houston vi har et problem."* Three sites in Denmark had deformed frogs also, the article reported, but an unconcerned Danish scientist sounded sure that the abnormalities resulted from bites by predators or were caused by parasites (Bech-Danielsen 1996).

Quotidien, a children's Paris-based French paper interviewed me and described the *déformations étranges chez des grenouilles américaines* in Minnesota (the strange deformities in American frogs), and said the *grenouilles malformées* (malformed frogs) are an *énigme* for scientists. Enigma was the right word (*Quotidien* 1996).

One news outlet created a false impression that cancer occurred all around the Ney Pond. An Associated Press reporter apparently overplayed a conversation with Cindy Reinitz, the teacher (Silver 1996). According to Cindy she was inaccurately quoted as saying there was "at least one person with cancer in every household around the wetland." On the East Coast, my brother saw the AP piece with a title something like "Ring of Cancer." A popular radio talk show host took it further: a fifth of the people in Henderson had cancer, he said.

I was flabbergasted by how rapidly a piece of potentially false information was disseminated by the media. Furious, the Henderson newspaper's editor sought a retraction from the AP, to no avail. In truth only two homes were located anywhere near the Ney Pond, and Henderson was miles away, across the Minnesota River. The editor feared a false cancer scare could damage the already struggling rural community.

For weeks during the fall of 1996, the publicity and media involvement took over my professional life and to a lesser degree my personal life, leaving me little time to prepare for our upcoming wedding or make space in the house for Verlyn's things. Some reporters even called me at home. One person, who sounded inebriated, called in the middle of the night. Finally, against the wishes of the agency's Public Information Office, I announced that I could no longer do interviews with the press. It had to be all or nothing because the MPCA media policy was to give access to all media, unselectively.

When a reporter asked, "When will you have some results?" I wanted to say, "Leave me alone so I can do the work!" With so much attention, we could not analyze our findings from the summer or work effectively with our research partners. We simply could not do scientific work if we were in the public eye full time.

I was glad when one TV show stated that it would take scientists at least five years to find the answer to the deformities, and I wished that everyone would see it that way. Most scientists—even the Canadians, who were a couple years ahead of us—knew that explanations for the frog deformities would not come any time soon.

That fall, the agency's water quality division staff assembled in the basement board room for the annual awards given to a few workers for various accomplishments, such as pressure under fire, humor in the workplace, or good customer service. Selected staff received formal-looking certificates they could hang on their walls. After all the official awards were given out (with applause), Mark and I were called forward. By then, staff had seen us appearing on numerous TV programs and quoted in many newspaper articles; we were quite worn down by it all. The director said something about the unusual work challenges we faced and proceeded to present us with an award of sorts: a smooth green ceramic frog.

One rear leg was missing.

I gazed at it in horror. The director smiled, but only a few staff chuckled. I felt embarrassed. We returned to our work area and promptly hid the shiny green creature out of sight. We never looked at it again. That "award" only symbolized to us the fringe status we had in our workplace; worse, it seemed to show a lack of empathetic engagement with the awful problem that we struggled to solve. This was not a joking matter. For Mark and me there was nothing remotely funny about the frogs.

After calling a halt to media interviews in the fall of 1996, I tried catching up at work.

One day I picked up the phone: "Judy, I have a message you should hear," said Tracy Fredin, who coordinated our Thousand Friends of Frogs educational program for school kids.

"Go ahead," I said, curious because he sounded perturbed.

"A guy e-mailed me about the 'huge infrastructure of institutions and scientists gearing up to focus the research about the deformed frogs only on pollution rather than other causes," said Tracy, obviously reading from his computer screen.

"He says that parasites are the only proven cause of multiple-legged frogs. He wants us to add a page to our website titled 'How Politics and Government Bend Scientific Research,' and he claims you've rejected parasites as a cause of the deformities because they're not pollution," he said.

"Oh, good grief," I replied. This was not the first time we'd been accused of this kind of bias. "If he'd called Mark or me"—he hadn't, obviously—"he'd have known we're already looking into parasites, predation, and diseases in addition to chemical causes. Who is he?"

"All I can see is his e-mail address—it ends in 'capitalism.com,'" Tracy replied. Much later, I looked up the Internet site and found a confusion of politically conservative publications about profit making, free trade, and individual property rights. We had already encountered negative attitudes from landowners' rights organizations, which had accused us of training "wetland vigilantes and citizen spies" in our volunteer monitoring project. Was that group driving this suggestion of "bending science"?

I talked with Steve Goldberg, one of three parasitologists examining abnormal frogs from Minnesota for associations between resting stages (called cysts) of a trematode parasite and the deformities. I asked what he was finding.

"So far, it's still pretty much across the boards," he replied. He found parasite cysts in both normal and abnormal frogs. Not only that, he sometimes saw cysts in a frog's normal leg but not its abnormal one. For some reason, Goldberg observed, the bones of deformed frogs seemed softer than the normal ones. He was skeptical that parasites would be the main cause of the deformities.

We both puzzled over the publicity that touted parasites as *the* cause of frog deformities.

In early November of 1996, my beloved Verlyn accompanied me down the aisle of my church (Falcon Heights United Church of Christ) to the sounds of gentle flute music interwoven with calls of loons and frogs. With work so crazy, we deferred our honeymoon until January, a trip to Mexico. By then, I hoped, things should have calmed down at the MPCA. With great effort and lots of give-aways we began to merge our two households into one and were more or less settled.

One morning I sat comfortably ensconced on the sofa while I journalled and sipped my second cup of strong coffee to rev up my engines before heading to work. Verlyn sauntered in, morning paper in hand. "There's something here you'll want to read," he said, handing me the front section flipped open to the commentary page.

The title jumped out: "State Agency Wants to Hear Only the PC Explanation of the Deformed Frogs." I jolted fully awake and grabbed the paper. The MPCA was accused of "dashing cold water on the theory of parasites as the cause of deformed frogs, because it was not politically correct," I read (Honsowetz 1996).

Stunned, I looked up at Verlyn and said, "This is awful. Whoever this is, he's calling the funding for the frog investigation 'a political boondoggle, typical of state agencies that want to protect their existence and staff.' I don't believe this! He's saying we *want* to blame human mistreatment of the environment and not look at natural causes, like the parasites," I said. "Where did he get this idea?" I asked.

I dressed rapidly and drove to work all riled up when the day had hardly begun. Not lost on me was the irony that the Minnesota Pollution Control Agency was being castigated for looking into pollution as a possible cause of frog deformities.

I rushed into our work area and handed Mark the article. He frowned as he read it. Exasperated I asked, "Is it my imagination, or does someone *not* want us investigating chemical causes?"

"It's beginning to look that way," said Mark, who reminded me of the accusation by agriculture interests that we were on a "witch hunt" when all we did was request a copy of the agriculture department lab's protocols they used to collect water samples for pesticide analysis.

"My dad just sent me this article," said Mark reaching to the top of one of his piles of paper almost as mountainous as mine. The biologist who touted parasites as the cause of deformities in frogs had sent an article to *Outdoor News,* a publication read primarily by hunters and fishermen in the Midwest, Mark said. I'd never heard of it.

In an article titled "Evidence Supports Theory That Frog Deformities Caused by Parasites," biologist Stan Sessions described his earlier work implanting resin beads at the base of limb buds in tadpoles and the multiple limbs that developed. A common parasite, not pollution, could explain the deformities, he said (Sessions 1996). The hypothesis was based on indirect evidence, namely that implanted beads, used as a surrogate for parasitic cysts, would block chemical signals to the limb bud and impair development. There was no direct evidence, then, obtained by exposing early frog larvae to parasites, to show whether the parasite could cause malformations.

"It sounds as if he's saying 'case closed,' " said Mark.

I agreed. "But what's this got to do with hunting and fishing groups?" I asked still baffled why Sessions was promoting his theory in *Outdoor News.*

"He says he's responding to a previous article that was skeptical about parasites as a cause of frog deformities," replied Mark. Later I dug up the article, in which Mark was quoted as saying there might be multiple causes of the malformations, and that parasites could not explain all the bizarre deformities we'd seen in the frogs, which was true.

Around this time a biologist named Charles Dailey had found young deformed bullfrogs at a pond in the historic Malakoff Diggins gold rush mining area in California. Most had extra limbs. Aware of Sessions's work with the implanted resin beads, Dailey told a news reporter in Sacramento that he'd ruled out contaminants from the mining activities because the pond was located *above* an old mine, not below it (*Sacramento Bee* 1996). His "seat of the pants" speculation was that parasites, not chemicals, caused the deformities. Later, Sessions did find parasitic cysts in some of the California frogs, but no pond samples or frog tissues were analyzed for the presence of toxic metals or other contaminants likely at a mining site. Just because the pond was above the mining site didn't mean the amphibians—which migrate over land and consume bugs and worms—were not exposed to heavy metals that moved through the food chain from mining-contaminated soils.

Shortly after Dailey's report to the Sacramento paper, a San Francisco paper ratcheted up the story: "Snake Parasite May Be behind Deformed Frogs." Heavy metals and other chemical causes were ruled out, the article said, because the pond was upslope from the mine site and the frogs didn't show the kind of neurological damage that heavy metals would have caused (Lehrman 1996).

"We've gone from a 'seat of the pants idea' to pinpointing the cause as parasites yet no chemicals were analyzed! What kind of science is that?" I asked Mark. It reminded me of a study of the deformed bullfrogs that were found in Iowa near a chemical plant. The biologist took no samples for chemical analysis, yet he concluded that parasites were the likely cause of the deformities there. "How can you rule something out when you haven't even looked for it?" I asked. Later, in a publication, the biologist acknowledged that they did not examine "numerous potential pollutants" and only looked at parasites in the bullfrogs (Christiansen and Feltman 2000).

Soon the news of Daily's findings jumped to national TV networks. Our Public Information Office staff person, Ralph Pribble, alerted me to watch the evening NBC news. I did, over a late supper. A teaser—"Mystery Solved"—ran, with a picture of a deformed frog. I laid my fork down. This was news to me. The newscaster listed several possible causes of the deformities that scientists had considered at the September meeting in Duluth—ultraviolet light, chemicals, or diseases. Then he intoned: "But a new study has ruled out those theories and pinpointed another culprit. California researchers believe a snake parasite may be causing the deformities."

"How can this possibly be?" I asked my nonscientist husband, who often had to listen sympathetically to my rantings about the politics that swirled around the deformed frogs. "I can't believe this! Charles Dailey's 'seat of the pants idea' is now proclaimed the gospel truth across the country!"

"This scares me," I said to Verlyn. "Look how easily the media has given this idea national prominence. Why couldn't they be more critical, or seek other viewpoints?"

A shiver of paranoia crept up my spine. Why were parasite proponents pushing their theory so hard? I didn't like the feeling. It was difficult enough to do the work before us, or at least try, without confronting such twisted reporting. Was some kind of behind-the-scenes effort trying to discredit attempts to investigate the frogs?

That fall, I was interviewed on National Public Radio's (NPR) weekly science program, *Living on Earth* (October 11, 1996). During the interview, I commented that I leaned toward chemical causes of deformities because, to date, the evidence supporting parasites by implanting resin beads in frogs was only indirect. To directly show that parasites could cause malformations, I said, tadpoles needed to be infected by the parasite itself, not just with resin beads, during very early development, and allowed to grow out the limbs. Doing this was the classic approach to establishing microbiological causes of diseases.

After my interview on NPR, Stan Sessions sent me a lengthy e-mail to correct "one important inaccuracy" in my NPR interview: the part where I had said there hadn't been a direct demonstration that parasites cause deformities.

My "more accurate" statement, Sessions wrote me, would be this: "We do not know what is causing deformities because analysis is incomplete. However, the only published experimental analysis of extensive limb deformities in natural populations of frogs shows that the cysts of parasitic flatworms can cause limb deformities in developing tadpoles." He wrote that mechanical blockage can cause deformities, a regulatory response that is featured "in every current textbook of developmental biology."

Sessions claimed the amputated and missing limbs, which comprised the majority of Minnesota frog deformities, were "ambiguous" deformities, because they could be caused by predation, lawn mowers, or maybe chemical trauma.

Ambiguous? Very few of the stumps or missing legs showed evidence of injury or trauma, according to the diagnostic experts who were carefully examining frogs from Minnesota. The cause of partial limbs was uncertain, that was true. To me such gross malformations were significant—there was nothing "ambiguous" about the severity of these misshapen limbs nor the effects they likely had on the ability of young frogs to move about and feed properly.

In his message, Sessions was coaching me word for word, to give *his* conclusions in my interviews without questioning them. *What kind of science was that?*

While his work with the resin beads did not eliminate the possibility that "something else is going on," Sessions continued in his long message to me, "It does show that it is not necessary for anything else to be involved." We shouldn't neglect the possibility that chemical pollutants might play a role, he said, and we should rule them out "if only to allay people's fears that humans might be harmed." But this approach "must be balanced by what we already know. Otherwise we are climbing a hill of gravel," he wrote.

Gravel hill, indeed. Mark and I were continually climbing a hill of gravel, one composed of our recalcitrant agency, confusing politics, and the many unknowns about the frogs. The message Sessions conveyed to me was this: focus only on parasites, don't waste your time and resources on chemicals, except, as he had put it, "to allay human fears."

One day, Ralph Pribble from our Public Information Office walked into our pod and greeted us by saying, "Mark, Judy, here's a couple of news articles from Missouri you might want to read."

Mark reached over and took one: it showed a boy with a deformed frog he'd captured somewhere in Missouri (Uhlenbrock 1996). By then, deformed frogs were showing up all over the United States, so this seemed barely newsworthy to us. Ralph handed me the other article from the *St. Louis Post-Dispatch,* titled "Deformed Frogs Caused by Fluke" (meaning parasites).

"Any suggestion that deformities were caused by pollution was misleading and inaccurate," I read. By now you'd think I'd be immune to this kind of biased message, but I wasn't. I gripped the paper, and started reading out loud to Mark and Ralph: "Nowhere in all the states and the provinces of Canada . . . has any evidence been found to implicate chemicals, pesticides, or man-made compounds as the cause of the deformities." Furthermore, it said, frog deformities have been found in Missouri for years. It's just the numbers that have increased. Biologists had discovered several years ago," the article said, "that a fluke, a parasite, caused the deformities" (Presley 1996).

I read out loud who the author was: Jerry Presley, the head of Missouri's Department of Conservation. "That's incredible!" I said. "It's as if the commissioner of Minnesota's Department of Natural Resources or the MPCA's commissioner had publicly proclaimed that he or she knew the cause of the deformities, even before anyone has fully investigated this."

"What closed the director's mind?" I wondered out loud.

The St. Louis newspaper article had quoted Missouri's amphibian expert,

Tom Johnson, who also worked in the state's Department of Conservation, as saying they'd had deformed frogs for some time in the state. I called him to chat about frogs. At first, he seemed to back the idea that deformed frogs had been around Missouri for a long time. But then he said, "We've recorded only fifteen deformed frogs."

"When were they found?" I asked, curious.

Twelve had been found that year (1996), one in 1993, one in 1992 and—he paused—"one in 1965," he said.

One deformed frog in 1965? To me this was scant basis for the claim that deformed frogs had been seen in Missouri *for years.* Was someone, perhaps his boss, pressuring him to downplay the state's more recent, small upsurge in deformed frogs?

It seemed as if an orchestrated campaign was underway, one aimed at stopping researchers from investigating chemical causes of deformities in frogs to focus only on parasites and "natural causes." Later, I accessed some of the articles written by conservative organizations in response to the deformed frog problem.

The Cato Institute, a nonprofit organization dedicated to "the principles of individual liberty, limited government, free markets, and peace" ran an article during that time titled "Environmental Science and Sound Science" (Gough 1998). In it, the author suggests that deformed frogs have been around for two hundred years and he implies that the increase in numbers of deformed frogs reported could result from more people out searching, a suggestion made by Stan Sessions (Lehrman 1996). Sessions's work, which hypothesized that the larvae of parasitic flatworms could cause deformities in frogs, was described as "a biologically sound explanation" and "good science," while the media, the Cato writer said, sees chemicals as evil and deformed frogs as casualties in "industry's war on nature."

One conservative newsletter ran an editorial titled "Leaping to Conclusions about Deformed Frogs" (D. T. Avery 1999a) in which I'm described as "undaunted" because I say that pesticides remain a hypothesis for causing frog deformities. If pesticides caused deformed frogs, then my government agency would accrue more staff and funding! We're accused of scaremongering, of making a political statement when nobody really cares about deformed frogs found by schoolchildren. The writer attacks the media for casting "eco-suspicion" about pesticide residues on fruits and claims that Rachel Carson was wrong when she said that pesticides can cause cancer. A cartoon accompanying the article pictured six crippled frogs (some legless, one with an eye patch) lined up to see a doctor. "Doc, what's causing our deformities?

Pesticides? depleted ozone?" "Nope," says the doctor, "You're all infested with natural parasites." To the side lurks a man whose briefcase says TRIAL LAWYERS. He asks if he can sue somebody.

In their book *Betrayal of Science and Reason* (1996), biologists Paul and Anne Ehrlich described the growing antiscience movement in the United States as "brownlash." This movement used code words like "sound science" and "balance" that usually meant giving credence and equal time to minority scientific findings that disagreed with well-established scientific research. A few contrarian scientists, for instance, would act as naysayers who would deny the reality of global warming, the health consequences from smoking, or harm from chemical pollution. They expected the media to give them equal time and attention. Later, the two-sentence Data Quality Act, passed by Congress in 2000, would legitimize in federal law the concept of giving "balance" primarily to industry-generated scientific findings (see epilogue).

Brownlash had simmered in the latter half of the twentieth century. An anti-intellectual movement was gaining strength, and scientific research projects were ridiculed and defunded. After President Reagan came into office in 1981, he removed the solar panels placed on the White House by former president Jimmy Carter and reduced funding for alternative energy. Creationists were succeeding in eliminating or crimping the teaching of the biological basis for evolution in public schools.

We had included investigations of chemical causes of frog deformities among other potential causes like parasites and ultraviolet light. Why some organizations saw this as unsound science baffled me. Was it only because a natural explanation for malformations, parasites, had been proposed? Why not keep exploring all hypotheses?

Why was Sessions pushing his parasite theory so aggressively? What prompted the leader of Missouri's Conservation Department to advocate so strongly for parasites as the only cause, not chemicals or other possibilities? Why did the Iowa biologist who studied deformed frogs near a chemical plant not insist on analyzing chemicals in frog habitats around the facility? Why didn't Charles Dailey have metals analyzed in that mining area pond out west? Why not consider alternatives?

We'd been accused of "bending science," yet we were exploring *all possible causes* of the deformities in frogs, parasites as well as chemicals.

Campaigns to discredit environmental science have been staged for many years by public relations firms hired by industry or industry-linked organizations. When scientists warned that global warming loomed as a real threat to earth, industry's strategy was to attack the science (Gelbspan 1997, 2004).

Decades earlier, chemical and agricultural interests waged an extensive media campaign against the EPA's proposed federal ban of the damaging insecticide DDT. They placing pro-DDT articles in newspapers all over the country. In one publication, a DDT supporter was shown eating the powder and smiling.

It seemed that *someone* was trying to bend science.

And it wasn't us.

CHAPTER 7

GOVERNMENT TO THE RESCUE?

After taking a week off for our wedding in early November of 1996, I returned to work with unrealistic hopes that the swirl of media requests and other communications might have simmered down. Instead, I was greeted by a flood of nine hundred e-mails and phone messages.

Cindy Reinitz reported on her students' latest survey of the frogs at the Ney Pond in mid-October: hundreds of frogs were still there, many with deformed legs and unusually tiny bodies, she wrote. They seemed weak and seemed to suffer from what Cindy called a failure-to-thrive syndrome. Also, some of the students had begun to question whether they were helping the frog investigation.

The students felt out of the loop, Cindy said. They didn't see how they could contribute next year, if at all. I affirmed to Cindy that the students' continued observations of the frogs at the Ney Pond were very important and helpful to us. By then our focus was statewide. I think the students really wanted to find the answers themselves and had come to realize, as I had, that this was not going to be easy. And the school lacked the tools required to analyze chemicals in the pond or in the tissues of the frogs.

Among the myriad of e-mails was a possible lifeline: a short communication from a director of a federal agency (the National Institutes of Environmental Health Sciences, or NIEHS), offering to help us in the frog investigation. Could they come to Minnesota to discuss this with the MPCA? My spirits soared because this agency, housed under the National Institutes of Health, worked to uncover connections between environmental contaminants and human health. We needed their expertise.

The NIEHS, started in 1966 by the US Surgeon General, works to understand how conditions in the environment affect the development and progression of certain diseases and other health problems in humans. Their research is broad, spanning lung diseases linked to smoking and air quality, mercury's effects on the fetus, the roles of diet and environmental chemicals in cancer, the effects of pollutants (such as PCBs and lead) on brain function in children, to name a few. The institute is located in Research Triangle Park in North Carolina, close to an EPA facility. The scientists interface with the National Toxicology Program (NTP), headquartered at the NIEHS. The NTP develops laboratory methods for testing and evaluating toxic chemicals and understanding the hazards that various substances in the environment pose to human health.

I continued scanning my messages. One from a reproductive geneticist at the NIEHS said simply, "Judy, I think we have a pretty good idea what may be causing the misshapen frogs. I am thinking about conducting a test in my lab. Give me a call when you get back to discuss." Could I send him a list of any chemicals we suspected for any reason as possible causes of frog deformities? he asked.

The prospect of expert help from an agency that grasped the wider environmental importance of the frog deformities revitalized us. Mark and I began to construct a list of chemicals to help narrow the field. Tens of thousands of chemicals are released into the environment. Besides widely used herbicides and insecticides and their additives, what others were likely in our frog habitats? Metals?

The paucity of basic research on chemicals or other agents that might cause abnormalities in amphibians limited our ability to construct a meaningful list of suspects. Some pesticides had showed up in tissues of a couple of frogs taken from the Ney Pond in 1995, and these went on the list. We added selenium, known to cause limb and eye deformities in wildlife in California, and other metals, including mercury, known to cause developmental defects. DDT, although banned, was still around and had been reported to cause kinks in tadpole tails. Methoprene, a mosquito control pesticide still in use, caused limb deformities in vertebrates, frogs included, as reported by one researcher at the meeting in Duluth. Any chemical in the environment that acted like retinoic acid, which participated in the early development of limbs, would be suspect. We added several types of newer pesticides, plus three herbicides we knew Don Ney had used around his pond.

I tackled more messages and returned a call from a farmer in Illinois. Since 1994 his pigs had been born with "tumors the size of a peach," he said. So many piglets died that he had to abandon his hog feeding operation. Frogs had come to his pond earlier and mated, he said, but they produced no young. Yet in the past he had seen thousands of little frogs. Biologists have noted reproductive failure among amphibians in some years, so perhaps this was not all that unusual. But the farmer went on to describe vegetation on his farm that had turned white. Most of his trees were dying, he said, because a pesticide had drifted over his land.

Was this a crank call or for real?

I called a staff person I trusted at the state Department of Agriculture about the whitened vegetation. He groaned. "I'll bet it was Command," he said. That summer, he'd spent considerable time chasing down calls about this herbicide. It had drifted onto people's properties and caused plants to whiten.

The Illinois farmer had suspected Command. The rest of his incredible stories could be true. Command went on the list of suspects.

One caller, who described himself as a "semi-retired consultant for a large pharmaceutical company," joked about people who lose their keys and insist on looking for them in the dark of night under the lamp post, where the light happens to be. "You can't discover the truth by looking only where you can see or only at what you already know," he said.

Where was this man headed? I wondered.

"I'm about to give you some free advice," he went on, perhaps sensing my impatience. "Are you aware of a class of chemicals called lathyrogens?" he asked. Some of them occur in relatives of peas, whose genus name is *Lathyrus,* he explained.

"Not really," I responded.

These chemicals cause characteristic lesions and deformities in bone development because they interfere with collagen formation, he said. Collagen, a fibrous connective tissue, strengthens limbs in early development before the bone solidifies with calcium.

My interest grew as he took me inside the nuts and bolts of bone development.

Rats fed lathyrogens have young with rear leg abnormalities, he said. Sometimes they have what's called the "broken back syndrome," because the spinal column is weakened. Cattle that ate a certain kind of vetch, a legume related to peas, during a drought in India produced calves with deformed bones, and people who ate the vetch seeds developed paralysis in their legs. Other research showed kinked backs, shortened legs, and stunted growth in toad tadpoles exposed to lathyrogens, he said.

By now, I was listening to every word this man said and scribbling down copious notes.

Immediately I searched the scientific literature and alerted the various researchers who were cooperating with us about this class of deforming chemicals. I mailed them packets of relevant scientific articles. Most exciting were some very sophisticated recent findings by researchers who had been able to pinpoint which particular part of known lathyrogenic molecules caused the limb deformities: chemical structures with names like organic nitriles and semicarbazides (Dawson et al. 1990, 2000, 2004).

Analytical chemists use a method called QSAR, for "quantitative structure-activity relationships," in which they examine chemicals known to cause biological damage to uncover which portion of the molecule's structure is responsible for the toxic activity. This can be very useful in helping predict whether other chemicals with similar structures might also be toxic. In the case of lathryogenic drugs, specific molecular structures were identified that, if found in other molecules, could cause deformities in bone. However, biochemists are well aware of a paradox: sometimes molecules that differ very slightly in structure differ completely in their biological effects. Sometimes even a minor change in the structure of a molecule *removes* its toxicity.

I added lathyrogens to our growing list of chemicals for the federal agency, and in this category I included an herbicide named cyanazine because it has one of the lathyrogenic structures (an organic nitrile) and was widely used in Minnesota in the early 1990s. Its breakdown products were found in groundwater in Minnesota (Thurman 2003), at the Ney Pond in low concentrations, and in rainfall at another study site. In my mind, it was a prime suspect.

Later, cyanazine was found to cause cancers and deformities in lab animals and banned from use (US EPA 2000) after more than two million acres of land had been treated in the United States with the herbicide, primarily for corn. Cyanazine was shown to cause a variety of birth defects in animals over a wide range of doses, and had low to moderate persistence from a few weeks to fourteen weeks, depending on the type of soil. It was frequently found in groundwater at low concentrations and was not broken down by water or the action of sunlight.

Our plans solidified for the visit of several scientists in December from the

NIEHS to the MPCA. I eagerly anticipated their arrival, as if they were the Coast Guard coming to rescue us from drowning in all the possibilities. Could they toss us the life line we so desperately needed? I admired this federal agency's aggressive actions to ferret out the truth: when a chemical company wouldn't divulge the composition of one of its pesticide products, a staff person had posed as a farmer and bought one hundred pounds of it. Then he analyzed the product's chemical components in his lab.

A foot of snow slammed down on the Twin Cities the day before the five scientists flew in to St. Paul from North Carolina, but it didn't disrupt their travel or our chance to meet. Their visit began downtown, where I had previously been scheduled to speak about the frogs to a coalition of dozens of environmental groups. Mike Shelby, the director of the federal environmental toxicology program, introduced his team: Jim Burkhart, who worked on environmental mutagenesis and who later became our contact person, plus a toxicologist, an analytical chemist, and a scientist who worked on environmentally induced changes in nervous systems. This high-powered group would connect us to basic research that might help unravel the causes of the mysterious frog deformities.

I welcomed their visit, though at the same time I had a deep sense of foreboding. How would my agency's foot-dragging managers react to these federal researchers, who seemed eager to explore the bigger picture of the frog problem?

We'd had frequent clues to our agency's resistant attitude. Just recently, Mark had asked for an extension of our much-needed student worker beyond the end of December, when our state grant for the frog work would run out. We could hear our section manager speaking loudly as he told our supervisor the agency would *not* extend our assistant. We suspected money wasn't the only issue. Something else was going on.

I could only speculate why the agency was making no effort to keep the investigation going. Did the managers believe that working on animals at the MPCA was inappropriate, that this belonged with the state's natural resources department, which dealt with fish, game, and endangered species? We'd encountered this attitude earlier, when we proposed using biological indicators to monitor streams and wetlands. Had management bought into the belief that parasites caused the frog deformities? Or did they fear that pesticides might be the ultimate cause and want to avoid locking horns with agricultural interests?

The most important part of that December day was a conference call that linked our five visitors with the other researchers who had been working with us. I wanted these scientists to hear what our collaborators had been learning about the frogs, even if the information was preliminary. We needed to share ideas and everything we knew at that point. That was the best way for science to work, I felt—with openness and questions, dialogue, shared knowledge, and trust.

First up was Roger Brannian, a histopathologist from the USGS National Wildlife Health Center lab in Madison, Wisconsin. He took us on a journey deep inside the bodies of deformed frogs.

"I'm seeing shortened legs that look like the bone didn't form properly," he said, his voice amplified through the speakerphone that sat in the middle of the table. Instinctively, we all leaned forward. "There's no evidence, such as inflammation, that these limbs were cut or injured," he said. He went on to describe the bones as thin and not calcified properly. They were primarily cartilage, or islands of bone surrounded by cartilage. "It looks to me as if the cartilage has failed to mature into bone," he said, calling it a "bone maturation defect." This fit with another biologist's observation that the bones of the deformed frogs were softer than those of normal frogs.

Brannian described deformities in the frogs' spinal columns, something we couldn't possibly have seen in our intact field specimens. I was fascinated by this inside view. "The vertebrae are misshapen," he said, "kind of like compression fractures." Some frogs lacked part of the notochord, which is a rod-like structure that helps organize the embryonic spinal column. I was excited by this, because notochord defects might be an indicator of a lathyrogenic chemical. He also saw some "corkscrew" crooked spines and some fused intervertebral joints. A few of the frogs had missing or reduced eyes, jaw deformities, and misshapen skulls, he said.

He had our complete attention. More than the legs were in trouble. The whole skeleton was involved. Until his report I had not known the extent of the internal deformities in bony structures.

I was grateful that this histopathologist spoke so openly about everything he'd seen in the frogs, that he didn't hold back. I wished all scientists could be so transparent and humble about their work.

Next topic for the conference call: parasites, which had received wide attention in the media that fall. Brannian continued: "I've found cysts of parasites present in some of the Minnesota frogs, but the cysts did not correlate with the limb deformities," he said. "The cysts are present in both normal and abnormal limbs. Some abnormal limbs have no cysts at all." He saw no evidence that parasites were associated with the other deformities in the spine, jaw, or eyes.

The parasitologist from California, Steve Goldberg, chimed in over the speaker phone. "I too have seen parasitic cysts distributed inconsistently in normal and abnormal frog limbs," he said. "Several deformed frogs that I have examined had no parasitic cysts in them." Goldberg sounded skeptical that the cysts caused the deformities. "It's not normal for parasitic infestations to fluctuate this widely from year to year," he said, reflecting what we knew by then: the deformities had surged globally during the mid-1990s.

The histopathologist from Madison agreed; he'd seen too many malformations not associated with parasites. "What I'm seeing in these frogs is more consistent with the actions of the lathyrogenic chemicals Judy told me about than with parasites," he said. "Lathyrogens can impede limb development and cause kinks in the spine in laboratory animals," he said.

"Perhaps we can lay to rest the parasite hypothesis," Jim Burkhart from the federal agency said, "and not invest any money in it." I disagreed, saying we needed to do more work on other frogs before we could rule out parasites as a possible cause, especially with all the recent publicity promoting them as the main cause of the deformities. These initial reports were only preliminary and we would want our research to hold up to scrutiny from all quarters.

At the end of the day, we filed into a spartan conference room on an upper floor at the MPCA to introduce the visiting scientists to some of our managers. Outside, we could see the white cover of snow over downtown St. Paul. We chatted briefly about Minnesota's winter weather. Then Mike Shelby, the national toxicology program director, took charge. Friendly and outgoing, Shelby introduced his staff, explained his agency's work, and outlined how they might be able to help investigate the deformed frogs, by analyzing water and mud samples from our frog ponds for chemicals, and by running lab tests on them.

"What we can't do is the fieldwork, such as collecting the samples or surveying the frogs," Shelby said, indicating that we needed to do that. Also, his federal agency would need a letter from the MPCA requesting help from the NIEHS before they could do any work in Minnesota, he said.

Our managers sat stiffly, listening in chilly silence. My worst fears were being realized. Mike Shelby's warmth could not melt the ice that permeated the room. I shivered, fearing that our bosses had already decided that the MPCA would not allow any more work on the frogs.

The meeting ended with a shaking of hands, but no solid assurances of a partnership.

Embarrassed by the unwelcoming reception we had given the visiting scientists, I went down with them to the front entrance. Outside the parking lot was plowed, but snow whipped around in the wind. As we approached the exit, one of the scientists, a man I was already getting to know as a straight shooter, said quietly to me: "I am absolutely stunned by the negative administrative climate you are dealing with here. They are not giving you squat," he said in low tones.

I thanked the scientists profusely and joked that they should come back in January and experience a *real* Minnesota winter. They laughed as they walked out into the brisk cold, happy to be headed home to the warmth of North Carolina.

In the elevator on the way back upstairs, a fellow staff scientist told me that the managers had just announced restrictions on out-of-state travel.

"Oh really?" I said, instantly worried about presentations I had scheduled at scientific meetings in Maryland and Texas during the coming year, and now an invitation from our visitors to come speak in North Carolina.

"Yes," he said. "I'm going to a meeting in North Dakota, but I'll have to stay overnight in northwestern Minnesota and drive across the state line before dawn to get there in time for the meeting." He grimaced.

We rode the elevator the rest of the way in silence.

Years later, I learned from a retired MPCA manager that the agency's upper level managers had been quite angry about the idea of accepting any assistance from the federal agency. The MPCA could take care of the problem by itself, they felt. He had told the other managers they did not understand how complicated the frog problem was, that it could take millions of dollars and years of work to figure out the cause of the deformities. The managers had "a microwave mentality," he said. "They thought you could just pop the problem in the oven for thirty seconds and then it would be done." Was that what they were thinking? Didn't they know that the causes of the majority of human birth defects were not understood? How could we quickly discover the causes of frog malformations given the difficulties uncovering those in humans?

The new year began with unexpected anxieties. At work, insanity seemed to be the norm at the start of 1997.

"Darn, another heavy breather," Mark said as he hung up the phone.

"Good grief," I asked, "how many *is* that now?"

"Seven." Mark had reported the problem to the state voice mail operator and requested caller ID, but we knew that was unlikely. "It's so weird, because whoever this is never says a word," said Mark, his voice troubled. Were Mark or his family threatened? He and his wife had two kids at home. Could this be some crazy person upset about our work on frogs or wetlands? We both felt intimidated by this anonymous breather.

Mark's phone rang again, startling us. "We'll be right down." He stood up. "Not the breather this time. It's worse. Dave says the freezer has warmed up. Things may be melting."

"Oh no, the frozen frogs!" I said, immediately thinking of the frogs that we'd carefully stored in the freezer, hoping we'd find the money to have their tissues analyzed for chemical pollutants. We sprang to action.

Mark raced down to the basement. I grabbed my phone and dialed Ice Man, our source for the dry ice we used to freeze deformed frogs in the field. It was almost four o'clock. I was pretty sure Ice Man closed then. A man picked up. "Thank God you're still there," I said, speaking fast. "We have a disaster here. Our frogs are thawing out. I have to get some dry ice right away. Could you please stay 'til I get there?" He assured me he would.

I flew downstairs to my car parked in the icy back lot, then roared out the rear gate and up the hill to Ice Man's facility, fortunately not far from the agency.

Never was I so glad to enter their chilly gray warehouse, where large insulated bins stored big blocks of dry ice. I almost hugged the guy who stayed after hours to help me. "Oh thank you," I said, breathless. "We have an emergency in the lab." Could he, I asked, slice around fifty pounds into inch-thick pieces so we could disperse the ice around our jars of frogs? He pulled on heavy gloves, hoisted a big block of smoking ice onto a table saw, and started slicing. The saw screamed. He wrapped the slices deftly in brown paper, then advised me to place the dry ice on top of the jars because the cold air flows down, not up.

"What is this for?" he asked, curious. I quickly explained our work on the deformed frogs, our melting collections, and our plans to analyze the frogs for pollutants that might have caused the deformities.

"Oh, I saw something about that on the TV," he said. "Wasn't that discovered by those kids?"

"Yes, there was a lot of coverage in the media," I said.

"Well, let me give you some scrap ice too," he said, walking over to a smaller bin. He filled a cardboard box full with smoking white scraps. It turned out we would need every bit he gave us.

I paid for the fifty pounds, knowing I wouldn't be reimbursed from work because agency rules precluded cash purchases within a fifty-mile radius of MPCA headquarters. Within that distance, orders had to be placed through the state's established contracts with certain vendors. It would have taken a couple of weeks or more to order dry ice that way. The small cost to me was nothing in comparison to what we might lose.

I returned to the lab and got a cart to load the boxes of dry ice from my car. Mark was moving jars from the failed freezer into three large coolers. "Some are melting," he said, sadly. "I sure hope this doesn't compromise the analysis."

My stomach lurched when I saw bloody fluid sloshing around the half-thawed carcasses of frogs in the glass jar Mark was holding. I suppressed an urge to scream. We saw a lot of hard work melting away before us. "Some of the sediment jars have broken," Mark said.

Feeling frantic, but working with care, we iced the half-thawed jars, and slipped broken jars of pond mud into clean plastic bags or wrapped them in foil.

I seethed. Unlike other labs where I had worked in the past, the MPCA had chosen not to install an alarm system or an exterior temperature recorder when purchasing the walk-in deep freeze. Too expensive to install the necessary wiring, someone said.

By the time we finished distributing the dry ice and had more or less rescued the frogs, a repair crew had arrived, looking rather bemused by our distraught demeanor. Somewhat relieved, we hauled the coolers into the freezer. We were about to leave when I remembered being short of breath once in a freezer because someone had left dry ice there. I grabbed a sheet of paper and posted a warning: CAUTION, DRY ICE INSIDE, VENTILATE FREEZER BEFORE ENTERING.

During this time, Mark and I struggled with how to handle a request from the Minneapolis *Star Tribune* newspaper to turn over to them *all* of our information related to the previous summer's deformed frog work—including site locations and descriptions of the frogs. I explained to the reporter why this was difficult and even risky for the research: people did not want the locations of their homes publicized. They feared loss of property value if it became known deformed frogs had been found on their land. The relationships we had with some landowners, jittery over what the frog deformities might indicate about their land, were already tenuous. If the paper publicized the locations, or reporters visited them, we could lose owners' permission to sample their ponds as important research sites in the coming summer.

Also, we didn't want casual collectors taking frogs. Recently, Bob McKinnell had told me about a commercial frogger out east who said he'd take any deformed frogs he could get his hands on.

The federal Freedom of Information Act (FOIA) requires government agencies to release records and materials to the public when requested. In a democracy, transparency of government's activities is of paramount importance. In the late 1990s the FOIA law was amended and passed with little debate to require that all data produced under federal grants be made available to anyone who wanted it. The National Academy of Sciences expressed concerns then about making data public even before the completion of a scientist's work. There'd be a "chilling effect" on research collaborators if lab notebooks, draft manuscripts, electronic mail, and raw research data had to be made public before analysis of the data and publication in scientific journals.

This is what we faced—having to give out our raw data, even before it had been proofed and analyzed. This potential for premature access to the frog data created tensions with our academic collaborators. They rightly wanted to review and publish their findings through the normal, scientific peer review process before their work got into the media.

The change to widen FOIA to force release of raw data before it was analyzed was largely driven by private sector interests that wanted full inspection of all data used to support government regulations. This would allow them to criticize findings in advance of the analysis and reporting. Scientists agreed that openness was valid, but leading scientific organizations raised the concern that releasing data prematurely could impair the scientific process—even politicize it. Finally, the act was revised to exclude preliminary analyses, trade secrets, copyrighted material, medical files, and drafts of scientific papers (Alberts 2005).

In spite of my pleas, the newspaper kept the pressure on and used the Freedom of Information Act to formally request *all* of our records. I knew they were entitled to access public information from governmental agencies, but release of site details and unproofed data could endanger the research.

I asked the reporter where the line was between thorough investigative reporting and actually doing scientific analysis. He replied that they "might find relationships we hadn't looked at." But could a nonscientist journalist do that? The newspaper was able to generate maps and create geographically linked information that might visually suggest false associations. For instance, printing a map that displays frequencies of both human birth defects and frog deformities, as one caller wanted to do, could have created a spurious correlation unless legitimate analysis had been done by a trained epidemiologist.

Sending out such raw data before we ourselves had time to carefully review and analyze it seemed out of bounds. The USGS didn't release environmental data until it was fully analyzed and written into a report. The USGS had some kind of protective firewall that we lacked. The MPCA either lacked such safeguards or was just too hungry for media attention to shield us.

The MPCA's Public Information Office (PIO) informs citizens and legislators about the agency's work, especially on issues that draw widespread interest like the deformed frogs. Our staff liaison with PIO, Ralph Pribble, helped us greatly by serving as a point person with the media, although often media people called us directly. He'd put out periodic updates on the frog work and organize press conferences. We needed Ralph's backup, but this time was different. He alluded to the $10,000 personal liability for state employees who withhold information requested by the public.

Occasionally, PIO sided more with the media's needs than ours, suggesting once that we should sacrifice one of our legally protected intensive study sites for media access or, absurdly, create a faux study pond we'd use to stage media events. As if we had time for that! One day, a summary I had written about an in-house planning meeting was immediately forwarded by PIO to a reporter, who called me the same day and revealed that he was privy to what I had just written. Shocked, I asked the staffer whether this meant we had reporters eavesdropping on *all* of our communications. He apologized. We were, sometimes, *too* transparent.

Finally, the newspaper compromised: they did not need the owners' names and phone numbers. With considerable relief, we deleted them. But they wanted the site locations. Mark scrambled to correct errors in the database, then sent it to the paper with more generalized locations. We knew a curious reporter could sleuth out the specifics of the properties and find the sites.

The climate at work grew more absurd. There was unrelenting external pressure on us to solve the frog deformities, yet our managers wanted us out of the whole mess. I knew we'd used up the state money allocated for 1996, obtained thanks to Representative Munger's efforts in the legislature the previous year. Absent new funding, how could we collect samples for our federal partner to analyze in 1997? The agency's budget priorities had long since been set. I was quite sure frogs were not included.

What was the problem? Were deformed frogs just too irregular, too much trouble? Did our leadership really not care? Were outside interests pressuring them to block a frog investigation? We were treading on fragile ground and only rarely dared to ask about future funding for the frog investigation. When we did, we were told that managers had to set priorities, that investigating the outbreak of deformed frogs had not been part of the longer-term planning process. I understood that, yet I kept hoping there could be some adjustment. We—and the agency—were in the spotlight on this issue, and citizens were asking for answers.

Mark and I felt trapped. The future for the frog research looked bleak. We couldn't continue to eat up our wetlands grants to pay for it.

"Without new support, I'm afraid we may have to shut down all the work we're doing on the frogs," I said to Mark in early January 1997. "Decline to give more talks, refuse to answer phone requests, not make plans for next summer's round of fieldwork. Don't analyze the frog data or write reports about them." I was more discouraged than I'd ever been, and I didn't like feeling that way.

"Perhaps you're right," Mark said, acutely aware that all the money for our salaries came entirely from our EPA wetlands grants. "We'll jeopardize the wetlands research if we keep this up," he said. Neither of us wanted to give up our primary work on wetlands. By then we had made good progress in refining our scoring systems that used aquatic invertebrates and wetlands plants, and our work was attracting interest from wetlands scientists in other states. We knew that our indexes, like a medical rating system that examined several indicators for gauging a patient's health, could be widely used in the future to document water pollution in wetlands.

I continued to refine and improve the invertebrate Index of Biological Integrity (IBI). The IBI was a sum of the scores for several measures of the aquatic invertebrate community: low scores were given if more pollution-tolerant species were present, like certain, but not all, kinds of leeches, and some beetles and bugs. High scores were assigned to samples that supported more kinds of dragonfly, mayfly, and caddis fly larvae, all known to depend on clean water.

The idea of walking away from the deformed frogs pained me deeply. If only our agency would *try* to help figure out the cause. By now frogs filled my thoughts and absorbed my life. But I had wetlands data I hadn't been able to analyze—counts of invertebrates by species to relate to chemical data from our wetlands sites, a report to write for the EPA on my wetlands work, a talk to prepare for a conference on developing biological criteria and scoring systems for wetlands. That was our future, if Mark and I had one.

One night that January I dreamed that Mark and I were at a legislative hearing. We conferred with each other and then I stood up and said we had used up all our money for the frog project. We were not going to work on frogs anymore.

I awoke feeling deeply sad. In truth, we had run out of money. We had to let the frog investigation go.

But I didn't want to give up. At least the public should be made aware that we had already consumed our special state grant for the frog work. I began to tell anyone who called that we had a gaping hole in the frog funding and this meant we would have to stop working on the deformed frogs.

Mark and I had already balked at our managers' suggestion that we ask EPA to convert our wetlands grants and use the federal dollars to fund the frog work. That would effectively kill our core work. The deformed frogs very likely pointed to problems in wetlands. But our scoring systems based on wetlands plants and invertebrates would, in the long run, provide a better way to rate pollution in wetlands.

Something had to change. I sent a lengthy memo to our management in which I made it clear that we had used up the special state grant. The federal agency (the NIEHS) could not analyze environmental samples unless we had crews on the ground in Minnesota. I summarized all possible options for funding the frog investigation from outside sources. But most were unlikely, I said. We had a problem.

From home, I called Char Brooker, a respected member of the Izaak Walton League, a national conservation organization that works to protect natural environments and includes members who hunt and fish and care about wetlands. I wanted her to know the realities. I mentioned the annual progress report I'd submitted to the legislature that showed how we'd used up all of the special funding for the frogs. I alluded to the lengthy memo I'd written to managers about the issue.

"Absent any new state funding, the frog investigation will cease," I said. She'd been completely unaware of this, she replied.

The next day, this ardent conservationist called me at work and formally requested a copy of the memo in which I had summarized the funding possibilities. Soon, the director of the Minnesota Environmental Partnership, a coalition of several dozen local and national environmental groups, with a goal of helping environmental organizations protect the state's natural resources, called and asked me to fax him the memo. Then Tom Meersman, the *Star Tribune*'s veteran environment reporter called, full of questions. He had already obtained a copy of my progress report on the frog grant, which was publicly accessible.

"Looks like you're out of money," he said.

"Yes, we are."

"Will the MPCA request more funding in its new budget?" he asked.

"I don't think so, but you need to talk with our section manager about that," I replied, hoping he'd tell the public the truth and raise pressure to force my agency to come around and keep supporting the frog investigation.

Meanwhile, plans were being made for a tour of the MPCA by state legislators on January 15, just before the start of the new legislative session when the agency's entire budget would come up for review. The agency wanted to make a good impression on the legislators who dealt with funding for environmental work by highlighting the MPCA's achievements. The House environment committee, chaired by Representative Munger, was scheduled to visit us in the morning, the Senate environment committee in the afternoon.

On display in the boardroom would be the live deformed frogs, the frogs that had been taken from us by PIO that fall. Unbeknownst to us, the *Star Tribune* planned to run an article exposing the MPCA's lack of future funding for the frog work on the same day the legislators would be visiting the agency. I should have expected something like this might happen, given my recent conversation with an active environmentalist who had plenty of contacts. But I was unprepared for the drama soon to play out.

On Wednesday, January 15, my day began with the usual cup of strong coffee and a brief review with my husband about our plans for the deferred honeymoon trip to Mexico at the end of the week. I grabbed the newspaper to scan it before heading in to work earlier than usual for the state legislators' visit to the MPCA.

I pulled out the local Metro section and froze: a bold headline, "State Plans to Reduce Research on Frogs" blared out at me. The Minnesota Pollution Control Agency would no longer research the deformed frogs, wrote Tom Meersman (Meersman 1997a), because the "problem is too large for them to handle, and needs to be handed over to federal authorities." There were too many sites with deformed frogs to come up with an answer to the deformities, our section manager had told the reporter. It would take an effort "beyond the mission of this agency," he'd said. There was more.

"This is awful," I said to Verlyn. It was seven a.m. I ran to the phone and called Mark at home to warn him about the article, relieved he hadn't yet left to catch his bus.

"Mark, there's a hair-raising article in the Trib," I said. "Meersman has revealed that we've used up all the money allocated for the frogs. He's quoted our boss saying that if the cause of the deformities is pesticides, the chemical industry would expect us to have that 'nailed down *nine ways to Sunday.*'"

I read Munger's comments in the article to Mark: "It's absolutely ridiculous for them [MPCA] to take that attitude. They were happy to get some research money last year, and now that there's a big problem, they don't want to be involved. I'm totally frustrated and disappointed in MPCA." The MPCA is "afraid they'll step on someone's toes, especially in the chemical industry. They don't want to bother themselves with anything that's going to antagonize industry," Munger said.

"Ouch," said Mark. "All hell's going to break loose on this one."

I hung up and flew out the door.

Driving fast, I tried to see the issue from our section manager's perspective. He was right. We didn't know what caused the deformities. Any proof that an agricultural chemical was the cause would certainly have to be rock solid, or "nine ways to Sunday," as he put it. But the way he'd said it made the agency look weak and ineffective.

Soon after Mark and I arrived at work, the House environment committee, chaired by Representative Munger, was bused to the MPCA from the State Capitol with our agency's commissioner on board.

On arrival, they were escorted down to the large meeting room on the lower level, where the agency's citizen board met, public hearings and press conferences were held, and staff had division meetings and holiday potlucks.

The tour, planned by the PIO, was aimed at convincing the legislators of the value of the MPCA's work while gaining favor for its upcoming budget requests. On display were live deformed frogs in an aquarium brought down from the information office and numerous newspaper articles. A TV monitor ran videos highlighting the diverse news coverage about the deformed frogs. One might think the agency planned to express the worthiness of the frog research to the legislators and then ask for funding to continue the work.

The legislators took their chairs in the boardroom to hear the MPCA's commissioner give his views on changing trends in air, water, and land issues facing the state. He explained how the MPCA was working to "balance the needs of constituents," by factoring in the "costs of environmental protections." To some of the staff, this attitude meant doing less to protect the environment in order to save industries some money.

MPCA staff spoke about leaking septic systems and unsewered communities, cars and air pollution, and animal feedlots.

Not one word was said about the deformed frogs.

After the presentations, the legislators were escorted into the biomonitoring lab, where other staff and I stood ready to explain briefly what we were doing to measure pollution in streams and wetlands using fish, invertebrates, and plant communities as biological indicators. I would talk about the wetlands work and briefly mention the frogs as a new kind of indicator of the environment. That was the plan, at least.

As our section manager came into the lab, he walked past me and said tersely, *"Don't you say anything."* I stood in shocked silence. Representative Munger glowered at our bosses from across the lab. I could tell he was profoundly angry at the agency for dodging the frog investigation, as that morning's paper had made clear. It pained me to see this man, whom I highly respected for his life's work protecting Minnesota's environments, now so upset and disappointed with my agency. I wanted to go over and talk with him. But I'd been muzzled.

As the legislators left the lab, one agency manager shook hands with my colleague, the fish biologist, then looked at me with anger and passed by in silence. That look alone told me I was headed for trouble. Clearly they suspected me of exposing the MPCA's hypocrisy about the frogs, and in a way I had, by telling the truth to outsiders about the lack of future funding.

People asked how I stayed at the agency given the resistance I experienced. It wasn't just that I needed the job. I was approaching sixty and could see retirement in the future. Both realities were true, but it was the core value of the work. I knew that I could never walk away from our wetlands research or from the frog investigation. Both projects were unfinished and too important for Mark and me. We'd invested years in trying to elevate the status of wetlands, in developing biological tools to monitor their water quality, and in the frog research. Neither of us ever thought seriously of leaving our jobs.

It was nineteen degrees below zero when I got up early the next day. The radio warned people to cover exposed flesh, because wind chills of sixty degrees below zero were expected. The predicted high for the day: six below.

I pushed open the icy storm door to grab the morning's newspaper for a quick review before heading to work. Oh to stay home today, I thought. Hunker down in this cozy house and rest. Maybe the car wouldn't start! Perhaps Verlyn was right after all to want to live in California.

I pulled out the local section, mostly to see the weather report. A headline, "MPCA Defends Pullback on Frog Research" glared out, along with my photo of a gruesome frog lying on its back with two pink-colored extra legs, folded at the knees, projecting uselessly up between the animal's two normal limbs (Meersman 1997b). The journalist had written a second story.

This time the MPCA's commissioner was quoted: "We are a regulatory institution, not a research institution," he said. He had told legislators that the MPCA would not seek additional funds or continue doing fieldwork on the frogs. The deformities were widespread in Minnesota and other states, and he felt that the research should be turned over to a federal agency and to academic institutions. "That's a university's job," he said.

Finally, it was out in full view, what I had suspected all along: the MPCA would have nothing more to do with deformed frogs.

A disappointed Representative Munger was quoted, saying, "I would think if the MPCA is interested in protecting the environment, they'd be out front forcing me to find the money, instead of me running after them and asking them to do something."

Rep. Jean Wagenius, a strong environmental legislator, expressed concern in the article for the schoolchildren who had discovered the deformed frogs. "Why are we telling them this is not important anymore?" she asked. "That's the message MPCA is giving them. 'Children, you gave us a problem and now we're going to back off.' That's real disturbing," said Wagenius (Meersman 1997b).

My husband tried to raise my spirits: "This might bring you some funding," he said gently.

"Either that or by the end of the day, my head could roll. Saying we don't do research is a slap in the face to all of the scientists at the agency," I said. I knew the managers were likely to hold me responsible for this embarrassment to the MPCA. My job classification was not secure. They could claim insubordination, as they apparently had for another agency biologist. He had raised a concern that mercury concentrations in the tissues of fish might increase if proposals went forward to create thirty or more new impoundments in the Red River basin in northwestern Minnesota. His bosses didn't go along with his expressed concern. Rather than engage in an extended and expensive legal battle to challenge the threat of his dismissal for insubordination, the employee retired early.

More recently, an MPCA hydrologist had been fired for undisclosed allegations about data handling when he had previously worked for the state's Department of Agriculture, testing streams for pesticides for sixteen years. He was placed on investigative leave after being asked to testify in the legislature about the behavior of atrazine in trout streams in southeastern Minnesota (McCormick 2007). He filed, then later withdrew, a whistleblower lawsuit (Meersman 2007a, b, c).

Arriving at work, I rode the elevator with a staff scientist. He looked at me wryly and said, "Well, I guess my work means nothing." His job was modeling pollutants flowing into rivers.

Another scientist came into our work area, drew a finger across his throat, and said, "Guess I'm out of a job. Could have fooled me, I thought my work meant something here." He had spent many years researching how phosphorus from runoff triggers destructive algae blooms in lakes.

The agency's expert scientist on mercury deposition from coal-fired power plants and other sources stopped by to empathize. "We should organize all the staff scientists to explain to management how the MPCA depends on their research, how science is essential for much of the work that's done here," he said.

In reality, science and research *are* necessary for any environmental agency to function. Well-founded analysis underpins all regulations and environmental monitoring. Staff scientists create the standards that support those efforts. They work on groundwater and sediment contamination, criteria for hazardous waste cleanups, ways to reduce mercury pollution and improve air quality. Their work protects both humans and natural ecosystems.

Working in the framework of federal laws, state scientists develop science-based regulations of pollution of air, soil, or water. State pollution agencies are charged with carrying out Clean Water Act responsibilities, such as overseeing the effluent released by wastewater treatment facilities and discharges from industries. This important federal law is administered by the US EPA, which has, in the past at least, provided states with chemical-specific guidance documents that help them set water quality standards, e.g., for certain toxic metals (Gross and Dodge 2005).

For the MPCA to develop water quality standards for a toxic chemical that harms aquatic life, for instance, agency scientists study peer-reviewed publications written by researchers from the EPA and academic labs. They judge carefully which levels of specific chemicals could be allowed in discharges into state waters without causing harm, then decide where to draw the regulatory lines. Before becoming legally adopted, any proposed standard goes through a process that includes public hearings and comment.

State waters are assigned "designated use classes." Public drinking water sources are given the highest use class and the greatest level of protection. Standards are set for many pollutants, for example, arsenic, toluene, vinyl chloride, or coliform bacteria, with the goal of protecting human health. Drinking water criteria for "Class 1" waters (in Minnesota) are often more stringent than standards for streams and lakes, or "Class 2" waters. Class 2 standards are aimed at protecting aquatic life, such as fish and other water organisms and plants, in addition to assuring safe swimming and recreational uses by humans.

The aquatic life standards provide the basis for monitoring streams and lakes, for finding and tracking sources of pollution, for imposing regulatory actions or fines, and for setting cleanup goals. Such standards can be used to trace pollutants from mining wastes, detect chemicals running off agricultural areas, and pinpoint discharges of improperly treated sewage or industrial waste. Rules are in place to define the number of violations of standards and the timing of samples needed to trip a regulatory action.

"It was disingenuous for the commissioner to emphasize our being a *regulatory* agency and to say that we don't do research," one scientist said to me. "All this time, he's the guy who has pushed hard for *less* regulation and more *voluntary* compliance by businesses. What does he mean we're 'regulatory'?" Someone suggested that the managers viewed scientists with skepticism because they might uncover too much evidence of pollution.

One staff person wondered if chemical industry organizations had pressured the governor and our commissioner to make the MPCA stop investigating the frog deformities for fear a chemical, like a pesticide, caused them.

Our phones kept ringing that day—people from environmental groups, federal scientists, and others who'd seen the newspaper article. Some were incredulous at what one person called the "stupidity" of the MPCA. Another claimed the agency's managers had "shot themselves in the foot."

I went down to the cafeteria to grab lunch, intending to bring it back to my desk so I could eat while I returned phone calls. Instead, I headed to a vacant spot by the window. As I passed a table, one staff person said firmly, "Keep up the fight!" By then, most of our colleagues knew about the challenges we had just trying to keep the frog investigation going. On the verge of tears, I didn't want my feelings to show. I moved away quickly, sat down, and looked out the window, barely able to eat. I was deeply disappointed by my agency and tired of trying to make them pay attention to what many people felt was a serious issue. I doubted I could rectify the situation, restore the frog work, keep up the fight.

The excruciating week ended on Friday with a meeting, the agenda for which was titled "Problem Solving Strategies." The director of the water quality division, a stocky man who wielded a lot of power, began by saying that he wanted to know how this blowout in the press had happened so we'd be sure it would "never, never, never happen again." I steeled myself to stay strong through the meeting. Might we be fired?

But the director's strategy surprised me: each person was given two minutes to speak, and Mark and I actually got to say something. The meeting was not accusatory, and we were not intimidated, as we had been in several punitive meetings we'd had that fall with our immediate bosses. Even our supervisor, normally critical, spoke supportively of the way Mark and I worked under fire.

To my amazement, as the meeting ended, the director suggested we might be given a full-time person plus half a student position to help us do fieldwork on frogs in the future. His proposal seemed unreal. Only a couple of weeks earlier they'd made us let go our student worker, who by then had valuable experience in managing the frog database and was knowledgeable about the field sites.

As we walked out of the meeting, our supervisor smiled and turned to

Mark and me and said, "It doesn't get better than this. Gee, I came in on my day off. I'm going to my boy's basketball game, and I don't know if the game is going to be as entertaining as this was." Was this just a game for him? A sport?

Mark and I shook our heads and smiled. Entertaining was not exactly the word we'd have chosen to describe what had just happened. Perhaps our supervisor was simply glad to hear the director's unexpected promise of new staff to assist the frog investigation.

On Saturday of that week Verlyn and I flew to Mexico City for our deferred honeymoon trip. I left work late on Friday feeling very skeptical about the director's almost schizophrenic offer of help on the frogs. In spite of what he said, I didn't trust him: it was a dangerous time for me to leave. He had to know that Mark and I were almost tapped out.

Our trip to Mexico affirmed my belief that effective government regulations are necessary to protect people and the environment from pollution. In Mexico City, lung and intestinal illnesses were common and linked to serious air and water pollution; you could see a photochemical smog build during the day, and the air quality was so bad that vehicles were assigned certain days of the week when they could be on the road. Tap water was not potable. I saw sewage running down a hillside in an open stream and children playing by the gray water. Except for Verlyn, most of our group succumbed to serious bronchitis and I spent much of the second week of our honeymoon coughing and recuperating at the beach condo.

I returned to work still recovering and not knowing what might have transpired during my absence. I entered a surreal climate where our section manager was unusually friendly, his voice unexpectedly calm. I knew he too had been under a lot of pressure because of the frogs. Now he was quietly asking me to construct some budget plans for frog work to have available for a meeting with Representative Munger. Was I dreaming?

I was galvanized into action, staying late at my computer to compose some rather large budget proposals to cover our outside research partners and fieldwork by an MPCA crew to begin in the summer of 1997. The only interruption was the security guard making his rounds and pausing for a brief chat. My energy surged at the prospect of continuing our investigation. At last the MPCA had turned around, even if against its will.

But then Mark and I were asked to meet in a conference room for a staff meeting led by the director of the water quality division. He began the meeting by saying, "Before Judy's trip you saw my good side. Now you'll see my bad side." I stiffened, not knowing what might come next. Around the table, everyone looked dead serious.

Then the director asked us, "How many of you consider yourselves deeply religious?" Mark and I and a couple of others hesitatingly started to raise our hands. "Well, good, because you're going to need it," he said, smirking. No one laughed. We squirmed in our chairs. He went on about all the "flak" that had come their way from the "frog thing" while I was gone and looked at me sharply, as if I were the cause. "I want you to know if we receive any money from the legislature for work on the frogs, it will have to flow right on through the agency to outside researchers. *We aren't going to use a penny of it,*" he said.

The urge to speak out against this ridiculous idea surged inside me. I bit my tongue and swallowed hard, aware that the director did not want one word from me. Once again the MPCA would try to have it both ways—getting credit for appearing to do something about the deformed frogs but in fact doing nothing.

Then the division director turned to me and said simply, "I don't trust you." My heart sank. "I haven't worked with you much," he said. "And right now my trust level is very low. This is a very sensitive time for this agency," he said. The biennial budget was about to go to the legislature, for one thing. Feeling completely humiliated, I girded myself, believing I was about to be fired.

I sat frozen, deeply hurt and with no idea how to defend myself. Never had I been accused by anyone of not being trustworthy. I had always tried to be straightforward and honest at work or anywhere. I was frequently the point person to the public at the MPCA for the frog problem. I had given numerous interviews with the media and never maligned the agency in any of that. I believed in the agency's basic mission: to restore and protect the environment.

True, I'd told an environmental leader about the lack of funding for continuing the frog work, but that *was* public information. Any good reporter could have found that out, as one had. And of course, when I called her, I had hoped that something would happen to rescue the frog investigation, soon to stall out. Even so, the fury that resulted from my "leak" of public information surprised me.

The director went on. "As researchers, you and Mark are busy counting trees and not seeing the forest," he said. I'd heard this stereotypic view of scientists before at the MPCA. In that view, scientists aren't able to see the "big picture." They are too narrowly focused on their own work to have any connection with policy issues, planning, or management decisions. Yet scientists are the ones who identify environmental problems, ask for solutions, and look into the future. The research of scientists underpins government regulations for human health and for a cleaner environment. What would be the consequences, for instance, of not using scientists' suggestions on how to slow global warming, of not cleaning up hazardous waste sites, of not reducing mercury emissions? Arguably, the "forest" for scientists differs greatly from that of managers.

Then the director turned to us, saying, "We are going to try to make you and Mark permanent employees." Surprised, our eyes flew wide open with sudden hope. For years we'd been working in temporary positions. We both wanted secure status at the agency and a guaranteed future there. Was he not going to fire us?

"But," he continued, "As newly hired staff, you both will be *on probation.* And you might be assigned to work you won't like. Up until now you and Mark have been doing pretty much what you want with your EPA wetlands grants. But that could change," he said. Mark and I knew that we had been fortunate in directing our work to what we (and the EPA) felt was important for protecting wetlands and their water quality.

"Of course, being on probation, you'll be the first to be laid off," the director went on. Then he paused and reflected. "Oh, no—you're unclassified now, so I guess you're already the first staff who would be laid off," he said.

Deflated, Mark and I left the conference room. We walked in silence back to our work area and sat down heavily. It felt as if we'd been fired. Yet we were being offered a secure position—only in jobs we didn't care about.

Soon after, my bosses and I drove over to the State Office Building, where legislators' offices were housed, to meet with Representative Munger about proposed budgets for the frog work. We gathered in a hearing room, where Munger spoke pointedly about his goal to keep the MPCA working on deformed frogs. I felt buoyed up by his impassioned pressure to force my agency back on track. Munger praised me, and said he'd like to funnel at least $300,000 to the MPCA for us to coordinate the research. If this passed the legislature, the money would become available July 1, the start of the fiscal year. Then we'd have sufficient funds to hire people to collect frogs and pond samples and to form contracts with outside labs for chemical analyses.

Meanwhile our section manager, still acting friendly, said he'd try to pull a few thousand dollars from existing MPCA monitoring budgets to support water sampling at frog ponds in the spring. We were interacting well. At last we seemed to be on our way, even if the managers were gritting their teeth and saying, "We *will* do frogs," when they wanted otherwise. My hope revived. We might get out there in spring after all and start working with our new federal partner!

But a sense of unreality hovered over me. I vacillated between thinking we'd lose our work to hoping we'd survive at the MPCA.

During that time, I dreamed I was planning a field trip and people were already waiting at the vehicles. But I was wandering around getting information we needed for directions. The library didn't have the kind of book I needed. The women there were dark-haired sorcerers, not like our friendly library staff. It was getting late in the day. People who knew me were coming up and saying, "There's something really awful; you had better lie low. It's a really difficult time." No one could, or would, tell me what it was. I had this feeling of dread that there was some big blowup coming because of something awful I had done. But I didn't know what it was.

I awoke, terribly perturbed about what had gone so wrong at work. Why couldn't I lie low, as a couple of colleagues had suggested. Let things play out?

Perhaps it was naive on my part to expect that the agency would willingly, even readily, adopt the frog problem or embrace our push to do biological monitoring to measure water quality in wetlands and streams. That was my "forest," my big picture but clearly not theirs. Managers of a politically structured government agency have a different sort of big picture from that of staff scientists. They have to take into account what their leaders—the governor, their commissioner—want; they have to worry about how the agency's actions might displease or satisfy the legislators who would vote on their overall budget; they have to respond to organizations of cities and counties, to environmental groups, to the Chamber of Commerce, the Corn Growers Association, the Pork Producers, and numerous other business interests in the state. Even managers who had a scientific background had to bow to political forces from outside the government and from further up the hierarchy.

The MPCA was particularly prone to outside pressures because it monitored and regulated pollution of water, air, and land. It could impose fines for violations of the rules. At the same time, beginning in the mid-1990s, the conservative Republicans in the US Congress worked actively with industry lobbyists to loosen government regulations. The effort to undermine the environmental research that supported regulatory criteria for pollution would only strengthen during the George W. Bush administration after the turn of the century. Similar pressures against regulations had to be in play at the state level. The immediacy of our battles at the time was so intense, I couldn't step back and think about how much and which outside groups might have been lobbying against the frog investigation or the wetlands research.

I knew the managers always struggled to gain sufficient funding from the legislature; that state agencies were rarely favored either by politicians or much of the public; that government couldn't do everything and its leaders had to prioritize the work. But did that explain the attitude our bosses seemed to show toward the frog work? Why had I not heard them say, even once, that they were troubled by what the frog deformities might indicate about the environment, even if they couldn't afford to investigate the problem?

I had always admired the Minnesota Pollution Control Agency for its early and dramatic history. I had wanted to work there long before I snagged that first temporary position in late 1989. In its early years, the agency had fought hard on many fronts for a cleaner environment, like its battle against Reserve Mining's enormous discharge of asbestos-laden waste into the pristine waters of Lake Superior. The agency restricted the dumping of hazardous waste where it harmed the environment and developed standards and regulations for toxic chemicals. Great strides were made during the 1970s and 1980s with new federal laws coming into being and the EPA working with states to regulate pollution. My section manager, there from the agency's beginning, told me about his efforts to convince towns to stop discharging raw sewage directly into streams or wetlands and helping them build wastewater treatment plants. "Anything you did made a positive difference in the environment back then," he said.

The MPCA staff had been my environmental heroes then; they still were, in spite of my disappointments. But something had changed at the agency during the 1990s, something I couldn't understand. As said, I'd been too immersed in work to step back and understand the broader implications of the efforts in Congress that strengthened during the mid-1990s to restrict if not undermine the science that supports regulations for a healthy environment.

During the winter and spring of 1997, I had no clear explanation for what actually drove the agency's management, only fragmented clues. For example, our bosses told us that the MPCA wanted the EPA to "loosen up the reins" on federal grants to the agency, so the agency's managers could redirect money to what *they* considered to be priority work in the state. A significant portion of the agency's regular budget consists of millions of dollars of grants from the EPA, grants that are targeted for activities such as wastewater treatment projects, controlling non–point source pollution, improving air and water quality, or cleaning up hazardous waste sites.

The idea of loosening the reins on EPA funds struck Mark and me with terror. We could lose all control over our federal wetlands grants. If given the chance, we knew that the MPCA managers would readily shift our federal money to other, higher priority projects.

Indeed, a couple of times that spring, our managers had already attempted to divert our federal grant money to inappropriate uses. In addition, after promising to the public that MPCA would provide the money to hire a person to work with Mark and me on the deformed frogs, the agency backpedaled. Our bosses sidled up and asked us to pay the promised new staff person out of our wetlands grants. We refused. That would have killed our core work.

When I argued this with our supervisor, he said, "Well, you brought this frog problem on yourselves, and you should pay the consequences out of your wetlands grants."

Mark made it clear to the managers that such misuse of our grants by the MPCA could be illegal and cause trouble if the EPA audited how our wetlands grants were spent. That got the attention of our bosses, and we succeeded in rescuing our money for its designated work.

That spring an article in the *New York Times* reported that state pollution control agency commissioners had sent the EPA a proposal that attempted to undermine federal regulations and mandates over states (Cushman 1997). Agencies like the MPCA are governed under federal environmental laws, such as the Clean Water Act and other laws related to the pollution of ground and air. Fortunately, the EPA saw through the ruse and blocked the proposed changes.

The proposal had purportedly been aimed at cleaning up "inefficiencies between states and EPA" and asked for more flexibility for how states spend federal dollars they receive from the EPA. But an EPA administrator saw it as allowing states "to circumvent virtually any regulation or statute." The lead author was none other than MPCA's own commissioner, Peder Larson. Unknown to agency staff, Larson was, apparently, working to undermine the legal framework that governed many of the MPCA actions that protected Minnesota's environments. This fit with what our bosses were saying at the time about wanting to "loosen EPA's grip" over the MPCA's activities.

In March of 1997 the pendulum swung in our favor—or that of the frogs. Under pressure, the agency had changed its mind (again), and decided that, should they receive any funding from the legislature to work on the frogs come July, *some* of it could be used in house. It didn't all have to "flow through" to outside researchers, as our director had previously insisted.

I was allowed to attend a two-day meeting of scientists to brainstorm about the deformed frogs in Madison, Wisconsin, in mid-March partly because the organizers paid my travel expenses. As I flew to Madison and was picked up at the small airport by a staff person from the federal wildlife health lab, my hopes revived.

In Madison, a couple dozen scientists convened in a large meeting room to discuss the frog issue. Joe Tietge, the EPA scientist from the lab in Duluth, led us through a process called "problem identification." The idea was to discover what we knew, what we didn't know, and how we might go about finding some answers about the causes of frog deformities. We divided into three discussion groups. I joined the toxicology and contaminants group, because it included some of the lead scientists who had uncovered the cause of deformities in fish-eating birds around the Great Lakes. I was extremely interested in how they demonstrated that PCBs (industrial oils) passing from fish to birds were the culprit. I was familiar with the framework they had used called "eco-epidemiology," but I wanted to hear how they applied it in solving the bird deformities.

The second discussion group focused on population and field issues for frogs, the third on disease development, or etiology. Included in that group were disease diagnostic experts from the federal wildlife health lab and an expert in viruses of bullfrogs, Dr. David Green, whose curled-up, handlebar moustache made him more look like a barbershop quartet singer than a scientist.

Stan Sessions was not present at the Madison meeting. Earlier, on a conference call organized by the EPA to plan work on deformed frogs in New England, Sessions had dominated the time. During that call he'd claimed, again, he that *all* of the deformities in Minnesota's frogs could be explained as caused only by parasites and predation. He'd described the missing limb and partial limbs as "minor abnormalities." The facile way he discounted the majority of the crippling malformations seen in Minnesota's frogs was simply shocking, but more troubling to me was Sessions's influence over the EPA's research planned for New England: parasites would be studied, but not chemical pollutants.

Hearing about Sessions's apparent attempts to thwart work on anything but parasites, a prominent Wisconsin biologist said angrily that he was being highly irresponsible. Work should continue on other possible causes.

At the meeting in Madison that March, the discussions among scientists were wide ranging. All the possible causes of deformities: toxic pollutants, parasites and diseases, ultraviolet light, and interactions among these, were discussed openly. In my group, Tim Kubiak was a USFWS scientist I saw as a bulldog when it came to fighting issues of toxic chemicals that harmed wildlife. Kubiak reviewed how he and others had researched the causes of deformities in colonial fish-eating birds—such as herring gulls, terns, and cormorants—around the Great Lakes. Birds with bill deformities, clubfeet, missing eyes, and defective feathers, along with slowed growth and higher mortality in their chicks, had been observed by bird-banders and by scientists.

Over years of research, the team attributed the abnormalities to industrial PCB oils and dioxin-like contaminants the birds consumed from fish (Kubiak et al. 1989; Gilbertson et al. 1991; Bowerman et al. 1994). Kubiak also pressed us to pursue endocrine-disrupting chemicals as possible causes of deformities in frogs. He talked with a sly smile that made him seem inscrutable, but what he said was clear and compelling. I listened, rapt.

I told our group about the pharmaceutical industry researcher who had called to tell me about the class of chemicals called lathyrogens and how they cause very specific changes in bones and the notochord. If he saw these flaws in frogs, he'd said, he could diagnose with certainty that a lathyrogenic chemical was the cause. This was the kind of diagnostic linkage we needed: specific kinds of deformities that might provide clues to specific chemicals or other types of causes.

At the Madison meeting, some tensions and disagreements arose among the scientists discussing field studies of deformed frogs. One group felt there should be a broad-scale survey of frog populations across a wide geographic area to deduce how much the deformities might harm overall frog populations.

"Who cares how many wetlands have deformed frogs," one biologist said, obviously riled up. We already know there are lots and lots of malformed frogs out there and in many locations, he said. Now we need to get at what's causing them. I chimed in, taking it a step further: "With human birth defects, we don't do research to discover how much those defects might affect human population size or structure, do we? Doesn't the research on birth defects focus on causes and prevention?"

The landscape ecologist countered, saying that their analysis might uncover associations between particular land uses and sites with deformed frogs.

I wished our bosses could have sat in on these meetings. I didn't think they understood how scientists deliberated, disagreed, and worked together to form hypotheses and develop approaches to solve problems—like frog deformities.

I left Madison to fly back to the Twin Cities filled with ideas and questions and empowered by talking with scientists who really cared about the deformed frogs. But as the plane landed, I wondered what lay ahead for Mark and me at the MPCA. Would we really be able to get out in the field that spring and coming summer to further investigate causes of the deformed frogs?

CHAPTER 8

THE QUEST

"Hey Judy, any chance your field crew could collect some fertilized frog eggs for us from one of your study sites?" asked EPA researcher Joe Tietge during a break at the amphibian meeting in Madison in the spring of 1997. He wanted to expose eggs to some chemicals and to ultraviolet light to see if deformities would develop, he explained.

I hesitated, still unsure if my agency would support any of our fieldwork that spring. "We'll try," I said. "But I can't guarantee we'll be able to get out in time."

Dave Hoppe, standing nearby, spoke up. "To find egg masses, you practically have to camp out at the wetland in May when the males are calling." And finding the eggs at a pond is a challenge because the frogs tend to breed in one local spot, not spread throughout a pond, he said. Our chances of collecting frog eggs for the EPA lab looked slim.

Back at work after the meeting, I sat down at my computer with a sense of urgency. I glanced out the window. It was late March, and I could almost hear the biological clock of the frogs ticking out there in the still frozen landscape. By May leopard frogs would be breeding and laying their eggs, exposing the vulnerable embryos to whatever harmful agents, chemical or biological, might be lurking in the water.

I turned to Mark, who was busily writing, and asked: "How can we get in the field this spring when the new funding, if it's approved, won't be available until July?"

With no real options, going to the ponds looked dicey.

Luckily, the MPCA held to its publicly made commitment to fund a "frog" position, and that allowed us to hire a person to coordinate fieldwork as part of an investigation into the frog deformities. I think the negative media exposure that embarrassed the agency in January had compelled our managers to do something about the frogs. At the last minute, right before frogs would begin to reproduce, we hired a young graduate student, Dorothy Bowers, who had experience surveying frogs across North Dakota and documenting spotted owls out west, to help plan and carry out the fieldwork. Joining us near the end of April with little time to prepare, she dove headlong into making plans for surveying frog populations, collecting and shipping samples to contract labs for analysis, and managing student field assistants. At first glance, one might not have realized that this diminutive woman with dark hair and eyes, whose small feet necessitated a special order for wading boots, had the take-charge and sometimes scrappy attitude that the unpredictability of the work required. Mark and I didn't need to warn her about the long work hours that lay ahead. She knew.

Even with Dorothy on board, Mark and I kept working double time to move the frog investigation and our wetlands research forward. Our bosses ratcheted up the pressure to complete the wetlands projects. Every two weeks, we wrote time-consuming, detailed progress reports for them, accounting for everything we did—or hadn't been able to do. No other staff had such close scrutiny. Our immediate supervisor and section manager wanted us to wrap up our federal grants and get them "off the books," as they put it. The biweekly meetings we had with them were punitive in tone, and we always went away feeling we had done something wrong. We were criticized for not getting enough done on our wetlands grants because of the work on the frogs, yet pressured to do more media interviews. After one especially difficult meeting, Mark said quietly, "I can't help but think our bosses are *trying* to make things unbearable for us. Are they increasing the pressure-screws on us to get us to quit our jobs?"

We were both at the breaking point.

Recently our supervisor, who hated doing paperwork, had asked me what my "best-case scenario" would be for hiring summer field staff, assuming Representative Munger's proposal to fund an investigation was approved. "Three," I said immediately, "One to work with Dorothy, two more as a second field crew to survey additional sites."

He leaned back in his chair and chuckled. "That's interesting," he said. "You and I see this in polar opposites. For me, the *worst*-case scenario would be having to hire three seasonals."

Our union representative believed we were being harassed during those work progress meetings. He suggested we file a grievance to try to stop the pressure-cooker sessions. But neither Mark nor I had the stomach or the time and energy to do that.

"You can't let it harm your health," the union rep had said. But that may have already happened. Mark looked pale and strained, and I couldn't sleep at night. At the meetings with our bosses my heart, normally steady, raced and skipped beats. I was short of breath.

When I described my stress symptoms to our supervisor, he simply said, "If it's too much, then you can quit your job." I clenched my fists, gritted my teeth, and walked away. We both knew I wouldn't do that. The work was too important for me to abandon. Or was it? Where was the limit?

"You'll burn out if you don't do something," Verlyn said one night after I'd unloaded some of the pressures I felt at work on him. "Seems to me your boss should be helping you, not making things harder." I nodded and took a sip of the wine he'd poured and wished we could escape somewhere far away and simply be together for awhile.

With me approaching sixty and Verlyn seventy-one, I contemplated retiring early. I created a spreadsheet to analyze my sources of income for retirement. A small but important annuity from my divorce settlement would help. But my Social Security record had way too many zero-input years, even when I'd had part-time jobs. I'd worked hard for much of my life but had very little to show for it. With only eight years in state government, my pension, based on the "average high five" salary years, would be minuscule. Then there was medical insurance. Leaving the state health plan would be costly. I just couldn't afford to retire. Not financially, not emotionally. And above all else, I had too much invested in our work to abandon it. It was part of my core.

At the last minute, the NIEHS, our new federal partner, threw another lifeline our way. They offered to fund the critical spring fieldwork, so they could start analyzing samples we'd collect from ponds known to have deformed frogs. They too wanted to analyze the water taken at the time the eggs were deposited. Luckily, our managers didn't refuse the money.

In early May of 1997 Mark, Dorothy and I drove away from the MPCA's warehouse in St. Paul and headed north to sample frog ponds. The rented van was packed full of waders, metal-handled nets for catching frogs, cartons of chemically clean jars, coolers for shipping water samples, maps, meters, cameras, and overnight gear. With Dorothy working out well under pressure, I felt we were headed in a good direction. We would be able to sample pond water at the most important time after all: when frogs laid their eggs. As we drove out of the city some of my chronic tension eased.

Our main quest was to collect water samples for analysis by the federal lab from ponds that had deformed frogs the previous year. If we happened onto eggs, we'd get some for the EPA researcher's experiments. We headed to hotspots in northwestern Minnesota first, sites where we had recorded high frequencies of deformed frogs the previous summer.

By early afternoon, we were driving down a county road through a rural area and pulled off onto a gravel drive that ran toward a trailer home. The owner came out smiling and greeted us, her two friendly black Labradors romping around, their tails wagging.

Downhill in the wetland we could hear frogs calling. My spirits lifted.

Armed with clean bottles for collecting water and garbed in chest-high waders, we stomped happily to the edge of the large wetland and waded in. I yelped and laughed when I stepped into an unexpected underwater beaver channel, my waders almost overtopping. We kidded each other. Freed from the poisonous atmosphere at work, Mark and I were actually having fun, something we rarely, if ever, experienced at the office.

Drawn by the grumbling calls of leopard frogs farther out in the wetland, we waded through the deeper open water to an isolated cluster of cattail clumps a distance from shore. There we stopped and stood transfixed. Directly in front of us, in a mere six-by-six-foot patch of cattail-sheltered water, were *dozens* of mating leopard frogs in a frenzy of mating. Hoppe was right. Leopard frogs *will* congregate and reproduce in a very small area. The exposure of eggs to pollutants or other agents could be very localized within a pond.

Some pairs of frogs, completely oblivious to our presence, were grasping each other in amplexus and about to extrude and fertilize eggs. Unattached males, their heads above water, called to attract a yet unpaired female.

I smiled at the scene, at this orgy of amphibian passion. It was a pleasant sixty degrees and sunny, a perfect temperature for both frogs and biologists.

We were on our way at last, collecting water precisely where the frogs were laying their eggs! We bent over and scooped up water samples near the mating frogs. Back at the van we labeled the bottles and iced them down in coolers, which we would ship as soon as possible to the federal lab.

Next we visited a large wetland behind the rural home of a Department of Natural Resources employee who worked in the agency's Detroit Lakes office in northwestern Minnesota. The family, whose daughters sold frogs to local fishermen, had observed only normal frogs around their property for the past few years. We hoped this would serve as a reference site to pair up with the hotspot we'd surveyed nearby.

Warned of an ornery ram, we trod carefully through the pasture behind the house and trekked downslope to a large wetland bordered by a dense cattail mat. We approached with caution. The owner had alerted us of deep water beyond the mat's edge. "If you step off, you'll be gone," he said. In most of our study ponds, we could wade through such cattail forests to get to open water. Not here.

The stand of cattails was so dense that we had to walk on top of the mat to get our water samples. I grabbed an old paddle lying on shore next to a dilapidated wooden canoe to use for support. Mark reminded me of one wetland we'd studied where the water was ten feet deep at the outer edge of the floating mat. Not knowing how deep it was, I moved cautiously onto the thickened mat, squishing it down with each step.

Mark and Dorothy slogged behind me. Suddenly, she yelped. One booted foot had punched through a weak spot in the mat. I looked back, and saw that Mark was helping her gain more solid footing. As I neared the water's edge, I leaned down on the paddle and crawled carefully toward the open water. I reached out and pushed the sample bottles under water and capped them. The mat's edge dipped a few inches into the water, but held. Then I backed up slowly and handed the bottles to Dorothy, now safely positioned on the mat closer to shore.

We drove on to another deformed frog hotspot we had surveyed the previous summer. The owner, who knew we were coming, came out to talk with us, his young daughters by his side, their lively Dalmatian scampering about. We could hear leopard frogs mumbling in the shallow wetland nearby.

I waded out. The water seemed strangely dark for such a sunny day.

All of a sudden I stopped in my tracks in knee-deep water, stunned by what I saw: a swarm of four-inch-diameter, black egg masses, so many they darkened the water for at least thirty feet around me.

"You won't believe this! I'm surrounded by frog eggs!" I yelled to Mark and Dorothy, who waded toward me. I estimated at least 100 masses floating close by. Within the past twenty-four hours, this meant 100 females had been in amplexus with males and extruded their eggs in this small, watery space. What an orgy that must have been! Later, Dave Hoppe told me his maximum count of individual masses had been 160.

My mind raced. Each mass contains 2,000 to 4,000 eggs. "There could be

400,000 eggs right here!" I exclaimed. I bent over to get a closer look. Already I could see the pale tan circle in the round black eggs that indicated development was underway.

We would have no problem here collecting the 400 or so eggs the EPA lab researchers in Duluth needed for their lab studies. It was almost too good to be true.

I squeezed off roughly a couple dozen jelly-covered eggs from each of several masses, slid the portions into clean glass jars with some site water and left an airspace. The remainder of each mass I let slither out of my hand into the water. As arranged, someone from the EPA lab would come to pick up the eggs. For them to carry out exposure tests during early development in the lab, timing was all important.

After we collected water samples from around the abundant egg masses and iced down the water bottles in our coolers, we drove away feeling exalted by our success. We had found frog eggs when Hoppe said it was hard to do, and we had collected water at the earliest, critical time for frog development!

We kidded each other and relished the idea of a late supper at the local sports cafe in the tiny town where we had a motel for the night. All was well, I thought happily. We were making good progress. Would the analysis of the water and eggs lead to solutions to the frog problem?

Suddenly we saw smoke. Dorothy swerved over and braked to a stop. We dashed out and raised the hood. Acrid smoke poured out. "I think it's the power steering fluid, Mark said, pointing to some liquid gushing out from a tube near the axle. Rocks and clay, the likely cause of the break, were packed in against the wheels. We drove into town without power steering, while I quietly fumed about not being allowed the use of sturdier field vehicles, like one of the agency's four-wheel-drive Suburbans.

The next morning, with the power steering line repaired, we drove in the rain to a site in a wildlife area whose clay access road was rutted and muddy. Recent heavy rains in northwestern Minnesota had saturated the land, causing major floods and evacuations of homes.

The van bogged down in the mud. We were stuck.

"This van has *no* clearance underneath and not enough power. It's made to drive to shopping malls and soccer games, not into wildlife areas," I grumbled.

In the rain, we stuffed cattails and pieces of cardboard torn from our sample boxes under the front wheels for traction. Mark and Dorothy pushed heroically, while I tried to drive us out, to no avail.

Dorothy charged off down the muddy road in the rain to find a farmer. Eventually they returned together in a pickup truck, but even that couldn't pull us out. The farmer went back for his tractor and finally pulled us to solid pavement. I offered, but he would not accept any money for his efforts. "Just glad to help," he said.

I got into the driver's seat and drove several hours nonstop to St. Paul. I had things to say to our bosses about vehicles. Mark and Dorothy—exhausted, wet, and muddy—quickly fell asleep.

The EPA researcher picked up the frog eggs for his experiments. The coolers full of iced water samples were flown east to the federal contract lab for chemical analysis.

Back at work the next day and still a bit groggy, I received word that the state legislature had approved the money for the MPCA to investigate the frogs: $200,000 for the next two years. Not the $300,000 that Representative Munger had wanted for us, but enough to get us into the field during the summer. The frog investigation had a future.

We got the go-ahead to hire summer workers to do field surveys of frogs and help collect and ship water samples. Our boss even agreed that we could access four-wheel-drive vehicles. No more vans.

I relaxed a little but wondered how long our luck would hold. From past experience I knew that I could never completely let my guard down.

Scientist Ed Little, who researches the biological damage caused by ultraviolet light and was experienced in measuring ultraviolet radiation in ponds, offered to help our investigation. He flew to the Twin Cities from the USGS Contaminants Lab in Columbia, Missouri, with his graduate student, Robin Hurtubise (today Calfee). I welcomed their arrival. Early on, some scientists at the EPA and others had suggested that certain wavelengths of ultraviolet light, known to be increasing because of the depletion in Earth's protective ozone layer, might trigger deformities in frogs or otherwise harm amphibian populations.

We talked about egg masses of leopard frogs that float near the water's surface. They could receive greater exposure to the sun's harmful rays than the eggs of other amphibians, such as salamanders, which are deposited in deeper water.

Dark-eyed and mellow-voiced, with a touch of a southern accent, Ed had a cheerfulness about him that belied the determined and dedicated scientist I knew he was. I suspected it would take a lot to rile him, and I doubted he had a grandstanding bone in his body.

I have a high respect for scientists at the USGS, housed under the US Department of the Interior. The USGS was first established in 1879 and is now a premier science organization that researches natural resources, investigates and evaluates wildlife issues, explores energy and mineral resources, analyzes chemicals in the nation's surface water and groundwater, and develops sophisticated mapping systems. USGS scientists work on climate change and acid rain; they document changes in snow and ice cover. Their biologists research biodiversity and ecosystems, wildlife health and disease, contaminants in animal tissues, surface water, and groundwater; others use sophisticated mapping techniques or track hydrologic processes of ground water systems. They gauge stream flow in rivers, Earth's geologic history, earthquakes and volcanoes, ocean processes—the list goes on. Scientists in the USGS, in my view, seem to have the least amount of political pressure and are most able to pre-review and publish solid scientific results before they go public.

On the way to our study ponds, Ed Little and I talked about the genetic damage that ultraviolet light causes to the DNA. UVB rays have wavelengths at the short end of the spectrum of light in the range of 290–320 nanometers and can cause skin cancers. But also, he explained, UVB radiation can transform certain water-born pollutants into more toxic chemical forms. The possibility of such photochemical changes makes it even more difficult to pinpoint which chemicals might be destructive to organisms once they get into the water. Ed and his team had discovered that UV light could alter petroleum pollutants, converting them into deadly chemicals that devoured fish tissues and increase the toxicity of certain pesticides (Little et al. 2000a, 2000b; Calfee et al. 2000; Zaga et al. 1998). These results had serious implications for humans, he said. Surgeons have been warned not to shine bright lights on a patient's liver during surgery for fear that contaminants sequestered in the liver could become toxic.

That the light from the sun, our fundamental source of food and life, the energy that drives our limbs and powers our brains, could change certain chemicals into more destructive forms made my skin crawl. It was bad enough that the increase in UV radiation is linked to a rising incidence in skin cancer.

Some amphibians, especially those species that lay their eggs near the water's surface, have evolved an enzyme, called photolyase, that prevents ultraviolet rays from mutating their genes. The enzyme repairs damage caused by UV rays to the DNA molecules. But other amphibians, like salamanders that lay their eggs in deep water, lack this protection or have less of it. Might the dark pigments in tadpole skin provide protection against UV? This was how our conversations went as we headed north.

Ed Little fostered creative thinking because he was not competitive and openly admitted what he didn't know. I liked that. Most scientists I worked with were similarly straightforward. But not all. Later, one researcher, who had never communicated about working with us on the frog investigation, unexpectedly did an end run around our loosely allied team of government and academic scientists. He wanted me to send him all of our research team's preserved frogs, but he hadn't asked any of us directly about this; instead he worked through another state agency, which passed his request on to me, thinking we had already refused him. After I confirmed that his proposed lab procedure would actually destroy the frogs' tissues, making them useless for any other kind of diagnosis, I communicated with this researcher by e-mail to explain why we couldn't give him our frogs right away. He called me and erupted in a temper tantrum, speaking so loudly and threateningly my colleagues stood by open-mouthed. He ranted on, barely hearing me explain that we needed to carry out more research on the frogs that we'd collected before their tissues could be dissolved. Once the other researchers completed their studies we'd try to work something out. But he wouldn't listen.

As we approached our first frog pond, Ed explained how he would measure light. The best data would be recorded during the two-hour period of maximum sunlight around solar noon, he said. He'd record the whole spectrum of light at three depths: at the water's surface, just beneath, and ten centimeters down, about four inches. "Below four inches there's usually a *big* extinction of UV," Ed said.

"What about vegetation?" I asked. "Often frogs lay their eggs in the weedy shallows."

"Oh, vegetation can also make a whopping difference in how much UV light would strike the eggs or tadpoles," he said.

I wondered how far ultraviolet light could penetrate into an egg mass, especially into the eggs that were positioned at the top of the mass. Would the jelly that surrounds them provide any protection? We chewed on this thought.

Summer days in Minnesota often begin with clear blue skies. As we arrived at our first site, clouds typical of midday were starting to build. Undaunted by the clouds, Ed stepped out of the van, looked up at the sky, and said cheerfully, "It's okay. As long as we can see the solar disc we can get a reading." I gazed up at the sky, skeptical. Smiling, Ed and Robin unloaded their equipment, worth at least $40,000, and calmly set up to take the first scan. A rare chance for me to observe how this was done, for them it seemed ordinary.

The spherical device for measuring light (a solar radiometer) looked like a five-inch-wide glass space vehicle mounted at the end of a six-foot-long handle. A quartz fiber-optic cable ran through the handle and connected to the recording equipment placed on shore. Ed waded in and positioned the radiometer just below the water's surface and stood holding the handle. Robin worked the computer, and they began. As he stood in the pond for the twenty-minute scan, Ed quietly asked me what I had thought about our media coverage, which opened up a whole new discussion. He was aware that some scientists were critical of the investigation, even going to the press with contrary results rather than sharing their findings with those of us doing the work. That government agencies or academic scientists competed with each other really bothered me, I said. And often it felt like the media put us on a public stage, not a good place for doing scientific research.

At a deeper wetland, one interspersed with emergent rushes and cattails, we carefully placed Ed's gear in an inflatable blue-and-orange float that looked like a small zodiac or life raft. We waded cautiously into thigh-deep water, each holding an edge of the raft to keep it stable for fear Ed's expensive gear might pitch over into the water. Fortunately the pond was calm and the wind was down. The water looked dark but clear.

We waded slowly, raft in hand, across the open water to the far side of the wetland where our field crew had earlier located leopard frog egg masses. We planned to document UV penetration at the same locations where frogs had recently laid eggs. I gazed across the water at the emergent plants fringing the pond's edge, at the scant border of trees near shore and the crop fields upslope. Behind us was a gravel road where we had parked our vehicle, near a home that faced a large lake not far from the pond. Its deeper water would be a good place for the frogs to spend the winter, I thought, an easy trek.

Suddenly, a bloated-looking tadpole leaped vertically through the surface. It was fat, shiny, and looked ready to transform into a four-legged frog. Maybe it already had popped its rear legs (which develop first). I couldn't tell.

Amazed, we watched, puzzled and transfixed. Another tadpole shot straight up, its mouth gaping. Then another vaulted into the air and slipped back into the water.

"What's going on?" I asked. I thought about tadpoles, how they breathe oxygen from water through their gills like fish, but then develop lungs as they mature into land-dwelling, air-breathing frogs.

We stood in the water in our chest-high waders while Ed started scanning the light. We speculated. Was the water too low in oxygen, causing them to gasp for air? "But we measured dissolved oxygen in the water, and it was fine," I said.

"Maybe the water's so toxic that they are trying to escape it?" someone asked. Later I came across an article that showed how excess lead contamination in water can cause tadpoles to leap, as if they were out of oxygen (Rice et al. 1999).

I remembered that a teacher at the Ney Pond had seen tadpoles leaping out of the water in June of 1995, near where the students first discovered the deformed frogs in August. The teacher had speculated that the water was too hot and the tadpoles needed air. A reasonable idea: hot water holds less oxygen than cold water.

Questions flew. Could the tadpoles be at the critical developmental moment when they convert from gill to air breathers? Could something be wrong with their gills so they couldn't get enough oxygen from the water? Were they exercising their new lungs?

"You know, acid rain releases aluminum complexes from soils into runoff to streams. The complexes clog the gills of fish," I said, remembering the photos of acid-altered fish gills, the fine filaments so thickened that water could barely flow between them. That's one way that acidified water kills fish, I said. Also, I'd read that an herbicide caused "erosion" in tadpole gills. Maybe something had made the gills less efficient at drawing oxygen from the water when the maturing tadpoles really needed air. We wished we could examine them. How were their gills? their new lungs?

Our speculations were wide ranging, but absent some experimental work, unanswerable. Did this indicate something bad in the water, or was it a natural behavior?

As we approached the worst of the deformed frog study sites, a pond located in north-central Minnesota, I updated Ed and Robin about Dave Hoppe's disastrous findings there: extremely high rates of deformities and grotesque malformations. In addition, many frogs and fish had died the previous summer in that pond.

After setting up their gear, Ed stepped into the pond with his UV wand, and they started taking readings.

"I can't believe this!" Ed exclaimed. "An awful lot of ultraviolet is getting into this water. It's at least ten times what we've seen penetrating ponds in the Rocky Mountains." This was an important result. In high altitudes mountain ponds are closer to the sun and the atmosphere is thinner. One would expect more UV light to penetrate.

I asked Ed to explain the difference.

"The Colorado ponds had a yellow tinge to them because of organic substances in the water," he said, explaining how the organic molecules effectively extinguish light. Our pond's water looked clear, likely because it was fed by underwater springs. Later our lab results confirmed that water from that pond was indeed low in UV-blocking organics. But it was the exception. Other Minnesota frog ponds where Ed measured much less UV penetration had higher levels of the dissolved organics, a more typical situation in Minnesota wetlands.

With each new finding, the frog mystery deepened. From Ed Little we learned that one pond's water—far more than that of the other ponds in our study—allowed an unexpected amount of ultraviolet light to penetrate. But at several other sites, UV was much less likely to be a problem for the frogs.

Was each pond in peril for different reasons? But if so, how would that mesh with the sudden rise in frog deformities in so many different locations in the 1990s?

Later, the impacts of UVB light on amphibians would be somewhat better understood (Little and Calfee 2010). UVB radiation in northern latitudes had increased 3 to 6 percent since 1980, with the highest amounts observed in spring, just when amphibians breed. But was that enough to harm their development and survival? Habitat quality—the presence of shade-providing vegetation and a greater amount of dissolved organic carbon in the water—could greatly reduce the amount of UV light exposure to developing amphibians. In lab tests, ultraviolet light would be shown to cause an increase in mortality in frog larvae, especially when in open sunlight (Tietge et al. 2001). Exposed larvae grew slower and were smaller at metamorphosis; their sun-struck skin could develop lesions that were subsequently invaded by pathogenic fungi. UV exposure caused bilateral limb deformities in leopard frogs (Ankley et al. 2000), meaning both right and left limbs were reduced. See Little and Calfee (2010) for additional citations of some work showing developmental abnormalities in UV-exposed amphibians.

In later years, species of amphibians would be shown to vary greatly in their protective responses to UVB damage (Little and Calfee 2010). Some have greater protective pigments in their skin, some are better at repairing the damage UV does to their genes. Behavioral adaptations vary: some tadpoles have the habit of hanging vertically or seeking shade. Some adults are more night active, others day active. Even the amount of protection from the jelly coat of egg masses varies by species. The increase in ultraviolet light certainly adds more stress on amphibian populations. But how large a role it plays in amphibian declines is still uncertain.

As I drove home after our field trip I thought how grateful I was for Ed Little's visit, for his open demeanor, for all he taught me. He brought a graciousness and light that eased my tensions about all the unknowns that faced us in the investigation. Who knew that UV could make less harmful chemicals more toxic?

CHAPTER 9

IMPERILED FROGS

Death and deformity again stalked Dave Hoppe's intensive study site in north central Minnesota in the summer of 1997. Frogs were floating, listless, and dying. Dozens of small fish and a couple of painted turtles had died. Young frogs couldn't swim or hop. Three-quarters of the mink frogs had severe and in some cases multiple deformities.

I read Dave's gruesome message summarizing this catastrophe and remembered the frogs I'd seen there the previous summer. Their bodies had been so grotesquely malformed I could barely stand to look at them. Some had had limbs projecting at odd angles from their abdomens where no limb belonged. In 1996 Dave logged deformed individuals in over 50 percent of the mink frogs, around 25 percent of the leopard frogs, and 4 percent of the toads. And this year it's *worse*? Seventy-five percent gone bad?

I told Mark the awful news. "It sounds like a frog Armageddon is going on up there." I couldn't imagine how the people who live by the pond feel, I said.

"They are worried," Mark said. He knew the family had been drinking bottled water, and they would no longer let their kids go in the pond.

I responded to Dave's dire message. Up until his report, we'd had many chilling experiences with the frogs, I said, but this was by far the worst.

Dave Hoppe had been very protective of that study site, the one where he'd found deformities in several species of frogs. Based on his observations there, he had proposed that frogs whose tadpoles live the longest in pond water, the mink frogs for example, would have the greatest number of deformities, compared with the toads that emerge early. Dave had exhorted us not to let our field crew collect tadpoles, as we planned to do at our other study sites so we could ship them to various labs. We had a team of academic and government researchers who were prepared to analyze the tadpoles' DNA and chromosomes for damage, to look for early-stage deformities, and examine them for evidence of parasites. Yet Dave had been quite firm that *only he* would collect from that site. He wanted to keep the frog populations intact and minimize any removals of frogs from the pond, he'd said. That meant our collaborators would not have tadpoles from that site.

But now, hearing his grim report, I implored Dave to let us at least collect some of the frogs to ship to the wildlife health lab in Madison, so they could diagnose the cause of death. We could take some of the dead ones, he said. But, I replied, they'd have to be shipped live, because after death animals can be invaded by opportunistic pathogens that might obscure the true causative agent.

Clearly disturbed by the nightmarish deformities and die-offs, Dave acquiesced. We could collect a limited number of moribund mink or green frogs, he said, as long as our field crew entered the pond by the dock, to avoid trampling any tiny toadlets that might be emerging from the pond along the shore.

That June, with less than two months on the job, Dorothy, our field coordinator, was out netting and examining tadpoles for early stage deformities from various ponds where populations of frogs had high frequencies of abnormalities the previous summer. Energetic and tenacious, Dorothy proved to be a great asset for Mark and me, a steady worker in spite of all the craziness at the MPCA and the ongoing bureaucratic hurdles we regularly faced and had to overcome. Coordinating the field work on frogs was at best intense, at worst chaotic, because our plans were necessarily always evolving. And throughout the frog investigation, the unpredictable seemed to happen with some regularity. There was no road map for anyone who headed into the jungle of the deformed frogs.

At one particularly stressful time in the field, Dorothy erupted and unleashed her frustrations at me over the difficulties of the work with us and with the agency. I could only listen to her pain. I couldn't fault her for the way she felt, as I was quite familiar with the challenges we all faced. Mark and I wondered quietly whether Dorothy might quit.

Tensions with our public information office, which was pressuring us to give more time to the media, increased in early summer of 1997. Our bosses created a kind of a "contract" that required us to spend defined blocks of our work time with the media and with William Souder, a writer who was working on a book about the scientists who were conducting research into potential causes of deformities in frogs (Souder 2000). Some of our MPCA colleagues were puzzled that management had imposed such an unprecedented requirement on us.

In addition to conducting interviews at our office, several news outlets and journalists wanted to go along with us as we surveyed frogs at various wetlands. In 1996 I had strongly resisted this, feeling it would interfere with the work. But now, in 1997, we had no choice. We agreed to stage three "media days" during the summer—at the Ney Pond south of St. Paul and at two ponds in the central and northwestern regions of the state.

On July 14 Dorothy and her crew went to Hoppe's study site to collect water for the federal lab out East and to find a few sick or dying frogs to ship live to the Madison lab for diagnosis. On her way back she called me, quite alarmed. "It is awful up there. We've got to do something!" She'd seen hundreds of dead and dying metamorphosing frogs, mostly mink frogs, she said.

Hoppe's earlier report, and now Dorothy's images of dead frogs floating near shore and listless frogs crawling out onto land, burned in my brain. What more *could* we do? There simply was no guidebook for this.

I called a researcher at the EPA lab in Duluth. Could his lab test water from this pond for acute and chronic toxicity? I asked. The lab's scientists were experienced in conducting toxicity tests on surface waters using minnows, small crustaceans, and algae. Historically, research done by the EPA lab had helped states set water quality standards for particular pollutants, mercury for example.

They could, he said. "Bring us at least ten gallons of water, chilled down to four degrees C." If I arrived at the EPA lab after hours, a guard could get the water into their walk-in cooler.

"By the way," he said." You know those eggs you collected for us in May? Some of the tadpoles developed kinked tails, and some of the metamorphs had enlarged lungs, at least four times normal size," he said. I knew some chemicals, like mercury, could cause kinked tails in frogs, but swollen lungs sounded really strange.

"What would cause the lungs to swell like that?" I asked. He wasn't sure. Afterward I talked with a federal scientist who'd found some evidence in the scientific literature that certain pesticides (in the carbamate class) could cause abnormally enlarged lungs. We had yet another mysterious symptom to ponder.

I learned later from a woman I knew at the Duluth lab that EPA staff had already carried out some preliminary tests on that pond's water, which they'd collected earlier from Hoppe's study site. "That's strange," I said to her, "He didn't say anything to me about this when I called." The puzzling lack of open communication contrasted with the free-ranging discussions I had been having with many other researchers.

"That water killed the amphipods (which are shrimplike crustaceans) within forty hours," she said. Hearing this, one of the MPCA scientists, who worked on toxic sediments that had accumulated in Duluth's harbor, commented that the pond sounded like a superfund site.

I called Jim Burkhart, our contact at the federal environmental health lab (NIEHS) and described the extreme conditions at Dave Hoppe's study site. He agreed to fund some additional tests exposing frog embryos to water from that pond.

The sooner we tested the water from Hoppe's site, the better. Whatever had so badly deformed the frogs at that pond and caused frogs and fish to die off could be breaking down or disappearing, if it hadn't already.

That July Dorothy and her crews were out on extended field trips. The only way to get pond water from Hoppe's site to the EPA in Duluth and to the NIEHS's contract lab in Oklahoma was to collect it myself. With help from MPCA staff, I found a dozen collapsible gallon-size plastic jugs to make clean water collections in bulk for the EPA. I loaded enough chemically clean glass jars to ship at least four liters of water to the Oklahoma lab, waders, cameras and several coolers containing bags of ice into an agency-owned Ford Escort. On a Friday in mid-July, I drove two and a half hours north to the pond.

The homeowner came out to greet me. "Nice day," she said, "but a bit on the warm side." I explained again what I'd be doing there. She indicated she was glad we were trying to understand what was causing the devastation in their pond.

"I sure wish I knew what this is," I said while booting up in chest-highs. I loaded a pack with plastic jugs and jars, and climbed down the steep hill to the pond. Its baffling beauty hid unknown and potentially lethal agents. To my left, a dairy farm bordered the shore. Behind the owner's property was a cornfield. Otherwise the site lacked major sources of pollution—hardly a place for a biological Armageddon.

Per Dave Hoppe's instructions, I waded into the water by the small dock to protect any toadlets that might be coming out of the water. I looked along the shore and remembered the past summer, when the ground had literally been carpeted with young leopard frogs, so many that we had had to go into the water to avoid stepping on them. The owner's dog had been having a heyday then, snatching up and eating little frogs.

I looked into the water and eyed a few dead frogs floating about near the surface, like abandoned bodies. By now they were haloed in a ghostly white fuzz, possibly caused by aquatic fungi or a funguslike infection, a protist named *Saprolegnia.* Those carcasses spooked me.

I waded in to fill the plastic jugs with water at intervals parallel to the shore. I tightened the cap and tossed each jug, which landed with a big splash, into shallower water. Then, huffing with the effort, I hauled them up the hill two at a time to ice down in coolers. Sweat ran down inside my waders and over my forehead onto my sunglasses.

The owner came down to help haul the water, but when I asked if she would hold a couple of deformed frogs while I photographed them, she balked. "I don't handle the frogs," she said. Normally, holding a frisky frog and taking its picture requires two people. But these frogs were so limp they lay passively on the grass. Spooked again, I shot a few pictures.

Exhaustion hit as I pulled off my chest-highs and loaded up the Ford. I glanced at my watch and panicked. "Oh my gosh," I said to the homeowner. "It's getting late. I have to make it to the Fedex office in Duluth before it closes at seven p.m. to ship some of the water samples to Oklahoma!"

"I doubt you'll make it there in time," she said, more familiar than I with the route north to Duluth.

I jumped into the car and roared off, driving faster than I had ever dared drive in my life. I pushed myself and the car, passing everywhere I could, knowing I was on the edge. Around quarter of seven, the Duluth harbor came into view, but I was not home free, because Fedex was on the north side of the city. I was south.

Thank goodness I'd written down directions to Fedex that morning. I whirled past the Miller Mall, past Gordy's Farm Market, and veered right when I saw the prison on my left. I raced down a barren road past the old airport to a small brick building. "This has to be Fedex, or I'm dead," I said to myself out loud. I pulled up, screeched to a halt and ran inside. It was seven p.m. on the nose. "Oh thank goodness you're here," I cried frantically to the lone woman inside. "I must ship water to Oklahoma tonight!"

"Take your time," she said calmly. "I'll wait for you."

I ran out and re-iced the cooler, strapped it up, and labeled it. I could have kissed the Fedex angel when I left to deliver the twelve jugs of water to the EPA lab located north of Duluth, close to the lake.

I pulled into EPA's parking lot and buzzed the guard, who, expecting me, stepped outside. "Thanks for being here!" I said.

"No problem," he replied.

"Is there a cart we can put these jugs on?" I asked. He rolled one outside, and helped load it, assuring me he'd get the jugs into the cooler right away. It was Friday evening. The EPA researcher had hoped they could set up the toxicity tests over the weekend. I drove out of the parking lot and gazed briefly at the twilight splendor settling in across Lake Superior. Relieved and tired, I turned south and drove home at a more normal speed.

Arriving late, I decided to return the state vehicle to St. Paul on Saturday and retrieve my car then. I had already planned to go in on the weekend to prepare my talk for the meetings to be held soon in Prague. I scrounged some food, briefed Verlyn on the day's events, then went to bed exhausted.

In early August, Verlyn and I flew overnight to Prague, where I had been invited to give a talk about Minnesota's deformed frogs at the Third World Congress of Herpetology. As arranged, we were transported from the airport in Prague to a starkly modern hotel built during the Communist era on the southern side of the city. The tall, glassy structure stood out in sharp contrast to the low, pastel-colored buildings in the old area of Prague, a city with a rich history dating back to medieval times, the city of Franz Kafka and poet-leader Václav Havel, the site of a World War II Nazi concentration camp. Verlyn anticipated delving into local history, especially of the Nazi occupation and the later "velvet revolution," when communist control was overcome without violence.

The meetings began with a plenary session in a large auditorium in the Congress Center, located a couple of stops on the Metro subway from our hotel. Following the conference opening, we dispersed for special sessions with speakers and attendees from all over the world.

I had given many presentations about both frogs and wetlands at public and scientific meetings. But this time I was nervous. I'd be speaking in front of an international audience of scientists. Would they understand me? Could I represent fairly the preliminary, sometimes sketchy, work of our researchers? Most of the biologists assembled in Prague had devoted their lives to understanding the natural history of reptiles and amphibians. Our backgrounds and languages may have differed, but we all shared deep concerns about threatened species.

I arrived early the day of my talk. I wanted to meet the leader of my session, an eminent British biologist, the now late Michael Lambert, who had invited me to speak at the conference. I also wanted to make sure that my slides could be displayed. The session, titled "Herpetofauna and Environmental Contaminants," would cover both amphibians and reptiles.

When I stood up to give my presentation, I saw at least 150 scientists in the audience, and no obvious media people. I began with slides showing the different types of deformities we'd observed in several species of frogs and toads in Minnesota. I saw people stop conversing and focus on my startling images. Next, I told how school students had discovered the frogs, and how I—and the state agency where I worked—became swept into investigating the problem. Feeling more at home, I summarized our work to date.

"I've just been told that one pond's water tested in the lab caused deformities in early-stage tadpoles," I said. Everyone looked at me, attentive. "This finding points to some causative agent in the water," I explained. "We don't know what that is, but chemical analysis is underway." In August we'd be testing water from additional frog ponds for effects on developing embryos, I said.

I was very pleased by the response to my talk. As I ended, scientists asked meaningful questions, and there was a buzz of concern in the hall afterward. I was approached by attendees from several countries: Australia, Denmark, Japan, Canada, and France. Each scientist wore a worried or puzzled expression, and several described deformed frogs they had seen in their countries. All wanted more information.

A Japanese scientist told me about deformed frogs that had been discovered in 1995 by a school student in one of their national parks in southern Japan. Until 1945 the land had been a Japanese military explosives site, the Yamada Ammunition Area in Kitakyushu City, he said. After that the US Army occupied the area until 1975. The site wasn't opened to the public until sometime in the 1980s, when it became a park. "In 1996, and again this year, over ten percent of the frogs we collected there were deformed," he said, in halting but understandable English.

Oddly, many of the Japanese frogs had extra *front* legs, a rare malformation in Minnesota frogs. Later Dr. Masayoshi Takeishi sent me published papers (Kadokami and Takeishi 1995; Takeishi 1996). Chemical analysis at the park showed high concentrations of DDT, the banned pesticide, but also benzopyrene and trinitrotoluene (TNT), a chemical left over from bomb manufacture. In the winter of 1999 Drs. Kadokami and Takeishi made a trip to the United States to meet with frog researchers and visited me at the MPCA. They gave us copies of their reports on deformed frogs in Japan. More recently, in 2010, Dr. Takeishi surveyed frogs along the Itabitsu River in Japan, finding 47 percent of 114 young *Rana rugosa* with partial and missing hind limbs (Takeishi 2011).

Over beer and hors d'oeuvres, I asked Leo Borkin, a Russian scientist, "Do you have any deformed frogs in Russia?" At first he said, "No." But when he finally understood me, he said, "Oh, of course. We have them around our nuclear power plants." He sounded as if that was not at all unusual.

Later he sent me copies of publications that detailed the extensive collections of frogs in areas of Russia that were polluted by discharge from paper-producing factories and municipal sewage plants. A startling 40 to 50 percent of the frogs had hind limb deformities and disrupted bony tissues, defects that were not cancer (Flax and Borkin 1997; Borkin and Pikulik 1986; Mizgireuv et al. 1984).

During the 1990s, the Russians had also studied frogs in an industrialized area of the Ukraine, where pollution from chemical and metallurgical factories and tire works was heavy. An astonishing 45 to *59 percent* of thousands of frogs they had collected were deformed. In their so-called better study area, where trees had been replanted and where ponds had formed in sunken areas over mining sites, 5 to 26 percent of the frogs were deformed.

At the rest of the sessions I listened enthralled to speakers whose stories and studies contributed to a widespread and disturbing picture of global harm to vulnerable species. Researchers from Denmark and France spoke of the loss of breeding ponds for amphibians. But this did not fully explain why their amphibians were on a downturn, they said. A scientist from New Zealand, Benjamin Bell, reported that three of his country's seven species of frogs had gone extinct, and two survived only on islands that remained free of invasive rats. A Swiss biologist, Kurt Grossenbacher, hypothesized that excess fertilizer running into frog ponds caused loss of oxygen in the water and this killed frog eggs.

Christine Bishop from Environment Canada began her talk to the hall filled with attentive scientists. "In Canada, the frog is a symbol of a clean environment," she said. She showed slides of promotional materials put out by Environment Canada. In one a frog says, "I don't pollute your home! Why do you pollute mine?" A chemical called rotenone, commonly used by fisheries biologists in Canada (and in Minnesota) to kill undesirable fish before restocking lakes with game fish, kills tadpoles, she said. Another chemical, used in streams to kill the immature stages of the noxious sea lamprey, a widespread pest of game fish in the Great Lakes, is also toxic to larval amphibians.

A British pathologist, Andrew Cunningham, described several diseases afflicting amphibians, including a kind of chytrid fungus, soon to be named *Batrachochytrium dendrobatidis,* or "Bd." The fungus invades and thickens the frog's skin, which reduces water and salt entry, affects the heart, causes lethargy, and often proves lethal to frogs and other amphibians. It infects the keratinized mouth parts of tadpoles, structures they need to scrape the algae coatings and plant material they eat. Chytrid fungus would soon be implicated in many, sudden losses of populations of frogs worldwide.

I listened to the presentations, heartfelt reports that suggested a dire future for frogs. I shifted uneasily in my chair, wishing I could go outside for a while, see the old city of Prague. The onslaught on amphibians was global in scale. The scope of this was far greater than Minnesota's deformed frogs.

Then Michael Tyler and Reinier Mann from Australia informed us that *twenty* species of frogs, which had been abundant just a few years earlier, were now considered endangered in Australia. "Something has changed in the 1990s, and we don't know what it was," Tyler said, his face troubled. In July of that year Australia had banned the use of seventy-four herbicides from areas near water because of threats to tadpoles and frogs, he said. We had no such comprehensive ban in Minnesota.

He went on. "Roundup, a common herbicide used for weed control both in Australia and in the US, caused die-offs of frogs." He'd tested the active ingredient (glyphosate) by itself for effects on frogs. Then he showed results from other tests where frogs had been exposed to the formulated product, the mixture as it's applied to fields. Other chemicals, sticky surfactants, for instance, are added to the glyphosate to make it adhere to leaves and work more effectively.

"The field mixture for Roundup was ten times more toxic to frogs than the glyphosate [the active ingredient] alone," he said.

I knew the use of Roundup was mushrooming in Minnesota, largely due to increased sales of "Roundup Ready" seeds that are genetically engineered to resist Roundup so the weeds are killed but not the corn and soybean seedlings. Later, researchers would confirm that formulated Roundup showed moderate to high toxicity to several species of larval frogs (Relyea and Jones 2009).

Several species of stream-dwelling frogs in Queensland, Australia, had a chilling status: "Disappeared from throughout its range." The declines of eight species were observed in a very short time period, from 1989 to 1994. One species, the northern tinker frog, disappeared in only two years (1989–91). Years later, only one of the eight, the green-eyed tree frog, would show any recovery from the declines seen in the early 1990s. And later, infections by the chytrid fungus were confirmed in frogs from Australia.

Similar declines were occurring in Brazil's Atlantic coastal forest (see Collins and Crump 2009; Alford 2010). In Colorado, the once-abundant boreal toad apparently began its decline in the late 1970s, and by the 1980s to early 1990s it was found in only one of hundreds of historic locations in the Rockies. Later work pointed to the chytrid fungus as the cause of die-offs of the boreal toad in Colorado. But why this infection surged was unclear.

A biologist told the dramatic story of the precipitous decline of the golden toad in Costa Rica and the reductions in many other species of frogs that began there during the 1980s. Over a few years of active searches, biologists documented the mysterious disappearance of this bright orange frog, which stood out in sharp contrast against the dark greens of the rain forest. Hearing the story of the desperate hunts for any surviving frogs moved everyone in the audience.

The golden toad disappeared rapidly after last being seen in the Monteverde Cloud Forest Reserve in northern Costa Rica in the late 1980s. At the same time, the harlequin toad and several other species had either disappeared or were rarely seen. By the early 1990s a disheartening twenty species of frogs in Monteverde had vanished since 1987 (Crump 2000; Collins and Crump 2009).

By the late 1990s the chytrid fungus was found in many dead and dying frogs in Costa Rica (Collins and Crump 2009). Could the increase in mountain clouds resulting from warming temperatures play a role in promoting the infections? Could wind-borne contaminants? Later research would show that some species of frogs seem more vulnerable to the chytrid infection than others. Some frogs produce less skin secretions that have antimicrobial activity, or they live at higher elevations where cooler climate conditions may have promoted the growth of the skin fungus.

But biologists were puzzled. Chytrid wasn't found in Europe and couldn't account for the diminished amphibian populations there. The chytrid fungus was present in Australia from the late 1970s, but not until the early 1990s were frogs found dead and dying with chytrid infections in Queensland. Why would this pathogen become more lethal around the early 1990s? What was compromising the frogs? This epidemic was unusual because the fungus attacked so many different species of amphibians.

Then a scientist from Colorado, Cindy Carey, stood up and explained how the sun's damaging UV rays might harm some species of amphibians more than others. "However, I don't believe that ultraviolet light is a major factor causing global amphibian declines," she concluded. Everyone's head went up. "Amphibian declines have increased suddenly in the past decade, whereas the amount of ultraviolet light striking earth has increased only gradually," she said.

The same is true for the deformed frogs, I thought. Deformities increased suddenly in the 1990s, not gradually over time. Parallel issues, parallel timeframes.

Gary Fellers, from the US Geological Survey in California, spoke with passion about the losses of frogs in the Sierra Madre mountains in California, and in the Sierra Nevada in California, where, he said, populations of five species of frogs were in a steep decline. Compared with historic location records for these frogs, a good 90 percent of their populations were gone. "The yellow-legged frog has disappeared from the western slopes of the Sierra Nevada. I attribute their loss to chronic exposures to pesticides blowing off California's heavily agricultural Central Valley," he said. Fellers described with visible dismay his failed attempts at recolonizing mountain streams that had lost their frogs, failures he attributed to pesticides. But game fish planted in high mountain ponds were also effectively eliminating other frogs in the Sierras, he said. Hungry fish readily ate the tadpoles.

This was not some dry, academic conference. Presenters conveyed intense emotions and showed concerns for these vulnerable animals. Amphibians were in danger all over the world. Were other forms of life in jeopardy?

Dan Pickford, a graduate student of Lou Guillette, a zoology professor in Florida, presented research from Guillette's lab which had been receiving widespread coverage in the United States, bringing attention to the role of environmental estrogens in the sexual development of aquatic species. Pickford's demeanor was serious, his slides disturbing. Young male alligators had small, deformed penises, and immature eggs were found growing inside their testes. The researchers' prime suspect: chemicals that mimic normal hormones such as estrogen. They suspected pesticides like DDT and other chemicals that had been spilled around a now-defunct pesticide-producing plant near Florida's Lake Apopka. These contaminants persisted in the lake's sediments and were incorporated into the food chain of the alligators, he said.

Even though I was sitting up front, I felt a stirring behind me and heard a murmuring among the scientists. Evidence kept piling up that both amphibians and reptiles were in deep, deep trouble.

In the 1990s, scientists began to express concerns that numerous chemicals used by humans in a variety of products and pesticides are released into the environment and disrupt normal hormone activities in both wildlife and humans. I'd read *Our Stolen Future,* published in 1996, with great interest. Dr. Theo Colborn had synthesized scientific data that supported the hypothesis that endocrine-disrupting chemicals could not only harm the reproductive organs of wildlife, they could harm humans (Colborn, Dumanoski, and Myers 1996). Colborn's book reviewed the early history of one of the first-known endocrine disrupters: diethylstilbestrol (DES), a drug given to pregnant women to prevent miscarriages and suppress lactation, a drug widely used in animal feed to promote growth in cattle and chickens. Called an "estrogen imposter," DES was linked to a rare vaginal cancer in young women,

"DES daughters" whose mothers had used the drug during pregnancy. Later, a wide array of chemicals was suspected or known to disrupt normal hormonal functions: other drugs, chemicals leaching from plastics, PCB oils, and some pesticides. Concerns would be raised: what role might hormonally disruptive chemicals play in breast cancer and declining sperm counts in humans? Research would broaden beyond concerns about chemicals that altered the sex steroid hormones of females and males to other hormones—thyroid and adrenal gland hormones, for instance.

Next, a lively professor from the University of California, Berkeley, Dr. Tyrone Hayes, engaged the audience with his easy smile and compelling talk. "In certain species, estrogenlike chemicals in the water will transform male tadpoles to females," he said. "Or these chemicals can delay the resorbtion of the tail during development, leading to giant tadpoles," he said while flipping on a slide of a bloated, oversize tadpole.

In Minnesota we had seen a few such oversize tadpoles that hadn't developed their legs properly. I had suspected some kind of interference with normal thyroid hormone, which is essential for the transformation, but I hadn't thought of a chemical mimicking estrogen. "Estrogen blocks the development of the thyroid gland," he said.

I listened intently while taking notes furiously. This got my attention. Could such hormonally disrupting chemicals cause deformed limbs?

Energized, and waving his arms as he pointed to his slides, Hayes finished by talking about DDT, the insecticide banned in the United States in the early 1970s but exported overseas by American companies long after the ban.

"Tadpoles in California have recently been found to have DDE [a byproduct of DDT] in their tissues, twenty years after its use was halted," Hayes said. He paused. "DDT makes the mouthparts of tadpoles fall off," he said. "It's been found in the eggs of frogs in Kenya," he continued.

Later Dr. Hayes demonstrated that atrazine, a widely used herbicide, causes ovarian tissue, or immature oocytes, to grow in the testes of male frogs. Since the conference in Prague, Tyrone Hayes has spoken widely about banning its use. The debate over atrazine and its regulation continues (see epilogue).

The conference concluded, and we packed up to head home after a day's delay because my passport was, apparently, stolen. On the long flight homeward to New York over the Atlantic from Amsterdam Verlyn quickly fell asleep. Never able to sleep on airplanes, I envied his easy repose. Instead, my mind churned with what might lie ahead back at work. What would the new lab tests show on the pond water I'd collected for the EPA and shipped to the Oklahoma lab? What had our field crews been finding in their frog surveys?

Just before I left work for the conference in Prague, a staff hydrologist had suggested that we sample the groundwater beneath the edge of the pond at Dave Hoppe's study site. Because the hill sloped so abruptly to the shore, the hydrologist suspected that ground water could be seeping into the shallow water where the frogs laid their eggs. What if the water under the ground was also toxic? Could the unknown cause of deaths and deformities in that pond be migrating there underground from a distant location?

My head was filled with everything I had heard at the meetings in Prague: problems for amphibians caused by pesticides; critical loss or degradation of habitats; disturbances from endocrine-disrupting chemicals; fungal diseases that might wipe out entire species; entire populations of frogs disappearing precipitously and mysteriously; deformed frogs discovered in several countries. The 1990s loomed as a dangerous time for amphibians, a time when frogs were threatened everywhere. The scientists assembled at the Third World Congress were clearly alarmed over this. I heard it in their voices. I saw it written on their faces. Their fear lodged inside my heart; it swirled in my brain. And it flew home with me.

"Judy, the water from two of your deformed frog ponds caused deformities in the lab frogs," Jim Burkhart from the NIEHS said after I returned from Prague. Early stage larvae had disrupted eye sockets and misshapen heads. Yet water from other ponds, whose frogs were all normal, did not trigger deformities in the lab's experiments. "There's something in the water. I'll fax you the results," he said.

I flipped through the pages of Jim's results, my excitement building. In these short-term lab tests, pond water caused malformations primarily in facial features and eyes of the larvae, but also in the gut. Moreover, extracts of the pond's mud had also caused deformities.

A bold graph made it starkly clear that *undiluted* pond water caused abnormalities in *100 percent* of the embryos. But when the pond water was diluted by half or more with clean water, they grew normally. At three-quarters pond water mixed with one-quarter clean, the numbers of defective larvae lay in between.

We had a dose-response relationship. The more the pond water was diluted, the fewer deformities it caused.

These were powerful results. We had evidence that both the surface water and mud from a couple of ponds could cause frog deformities in the lab.

I called him back. "Your results suggest that we can set aside parasites and UV as direct causes, at least in these ponds," I said. We agreed: it was most likely that some chemicals in the water and mud were causing the deformities.

"There's more," Jim said, almost whispering. "I've just heard that the shallow groundwater your hydrologist collected at Hoppe's site also caused abnormalities."

I paused. Whatever it was, it was everywhere. Even beneath the ground.

"We need to investigate the deeper groundwater," I replied, wondering how we could pull this off.

Later that day I talked with our hydrologist, Joe Magner. "The easiest way to do that," he said, "is to draw ground water from existing wells." We agreed that he or Chuck Regan, another staff hydrologist, would collect water from wells of the few families who lived close to three of our study ponds, then ship it to the lab for the embryo assay.

One evening Jim Burkhart called me at home, something he had never done before. I went on high alert.

"Judy, we have positive results on the well water assays," he said, his voice low and grave.

"Water from all three wells caused deformed frog embryos in the lab test," he said. Jim's contract lab employed a standardized test using fertilized eggs of the African clawed frog, *Xenopus laevis.*

I spaced out momentarily, barely grasping the implications. The homeowners' drinking water caused deformities in the frogs? "Jim, this is really bad news," I said. "We'll have to let everyone know, but how?"

We paused, thinking.

Jim would soon fly to Minnesota for a previously scheduled meeting in early September to discuss the frog investigation with representatives from other agencies. Now, considering the tap water results, we stared at something truly frightening: a public health threat.

I wondered where this new evidence might lead.

First thing the following morning I told our section manager about Jim's troubling report. Immediately, he ordered MPCA staff hydrologists to head north and collect tap water from the homes and deliver it to the state Department of Health. Their lab would run routine analysis for dozens of chemicals to see if established drinking water standards were met. Also, he ordered that the families be supplied bottled water as soon as possible.

The next day Jim Burkhart and a representative from the federal Centers for Disease Control arrived to meet with the MPCA and representatives from the state's health and agriculture departments. Jim presented the latest results from the frog embryo tests.

"There's a concordance," he said as he flashed data on a screen. "Water from locations having deformed frogs caused deformities in the lab, and water from normal ponds did not." They don't know whether limb abnormalities would develop, because the standard lab test runs only four days, he explained. A longer-term assay is in the works.

"But there's more."

Everyone's attention was riveted on Jim as he reported that well water from three homes caused larval deformities in the lab test.

People stared in disbelief and turned to each other, their lips tightening, their eyes widening. Drinking water?

Emotions ran high and discussion erupted around the table. What are the implications of these results? Is the frog assay relevant for humans? How to proceed?

Jim wanted to sample many more wells from homes close to ponds with deformed frogs.

Then he explained how they would analyze the water to sleuth out its chemical composition using a fractionation process. Water is passed through porous vertical columns of materials, such as activated charcoal or a material that separates molecules based on their size and weight. The water that passes on through the column would be tested to see if it caused deformities. If it did not, then the chemical agent was bound up on the material in the column. These were important first steps in isolating and then identifying what the chemical was.

We suggested that MPCA hydrologists work with USGS scientists to understand the groundwater. Was it, God forbid, flowing toward larger aquifers used for municipal drinking water supplies?

A big issue was how to tell the public. One person argued for solidifying these preliminary results before letting them out. But the MPCA's director of public information pushed strongly to hold a press conference as soon as possible. I agreed, as did the representatives from the state and federal agencies.

"Better the information come directly from us. We can put a bracket around this thing, so it doesn't blow up," she said. "We have to let people know; otherwise there could be a widespread panic about the drinking water." She suspected that rumors were already flying around the local area and someone might contact the press.

Weeks before this, ABC's popular news commentary program, *Nightline,* hosted by Ted Koppel, had scheduled an interview with me to take place the next day. Our assistant commissioner said I should be frank about the well testing results and tell them the state is providing owners with bottled water. But I could do the interview only if the TV network would hold it back until after our press conference. I contacted the *Nightline* producer, and he agreed.

We hustled to clarify our messages. The embryo assay was short term, typically run four days; it tested developing frogs, not mammals. But our federal partner had told us that the frog test had a high correlation with developmental abnormalities in rodents, and it was being standardized as a tool to screen chemicals that might cause human birth defects.

"It looks to me like parasites and ultraviolet light are ruled out," said a toxicologist from the Health Department. That leaves chemicals and maybe some disease agent. If it's a chemical, then the water *could* affect human health, he reasoned. "We'll just have to say it's too early to know what this is or what it might do."

To help us prepare for the press conference, we generated a list of questions we might be asked:

- Will all people who have deformed frogs on or near their property receive bottled water?
- Will the state test wells around all the sites that have deformed frogs?
- Why didn't you get at this sooner?
- What contaminant is in the water?
- What's causing the deformities?

On the day of the press conference, held in the MPCA boardroom, dozens of TV newscasters and print reporters flooded the room. I sat tensely at a long table with the MPCA commissioner, Peder Larson; George Lucier, the director of the Environmental and National Toxicology Program; and Jim Burkhart. We faced a bank of video cameras and media people.

Commissioner Larson led off by reading from the agency's official news release. "Our new findings," he said, "may provide a significant step toward understanding why so many frogs have turned up deformed in Minnesota and elsewhere in recent years."

Speaking next and summarizing the results that Jim had shown us earlier, Dr. Lucier concluded that we don't know what the causal agent is, and we don't know if the water could harm humans.

"However," Dr. Lucier said, "there is basic agreement between the frog embryo assay and rodent bioassays for birth defects."

The room grew silent.

Pens scribbled and little was said.

I liked Dr. Lucier's clearly honest and open statements, his expertise and unflappable manner. Next he spelled out his agency's plans to analyze the pond water and to host a workshop at their facility in North Carolina in December for scientific discussion of the malformed frogs. He opened up for questions, which flew at us from the crowd of media people pressing in near our table. TV cameras swung to each responder and recorders whirred, as we replied as best we could.

At the end of the press conference, we got up to leave. A reporter from the back zinged a question at me, one we hadn't rehearsed.

"Dr. Helgen, would you drink the water?" he asked.

"No, I wouldn't, not until I knew what's in it," I replied, being honest.

For a few days after the press conference, my supervisor kept chewing me out for giving that answer because it had made the news. I had to be truthful. What else could I say?

That I knew of, drinking water that's processed and ready for distribution to the public is rarely if ever tested to see whether it could cause deformities in lab animals. Of course the water supply is analyzed for many chemicals and for bacteria that might have health effects in humans. The legal standards for allowable concentrations of chemicals and bacteria in drinking water are derived from scientific toxicity tests and are set at levels to minimize harm to humans. Public water supplies are regularly monitored. Private wells are not, unless the home owner elects to pay a testing lab to check their water.

Immediately following the press conference, we were flooded with phone calls. In addition to personnel in the information office, the MPCA added a team of four staff responders to help field the calls. The commissioner went on CNN and explained the importance of a government agency being open with the public.

Don Shelby, a well-known local TV network newscaster, called me. He didn't want to play up a potential danger with drinking water, he said, and unnecessarily scare people. I told him that our data were limited and preliminary, that we didn't know if the results with frogs indicated that any harm might come to humans. On the evening news he didn't sensationalize our findings, which I appreciated.

The *Star Tribune*'s cartoonist, Steve Sack, depicted Kermit the frog explaining to a disgruntled Miss Piggy why his three-legged, three-armed, flap-capped uncle had arrived with suitcases. Pointing at him, Kermit says, "My Uncle Ole from Minnesota. He'll be staying with us 'til they get his well-water problem fixed back home" (Sack 1997). The paper also ran an article, "Private Wells Linked to Frog Deformities" (Meersman 1997).

Stan Sessions, the outspoken proponent of parasites as the most likely cause of frog deformities, immediately jumped into the news. Sessions said he was "absolutely astonished by that press conference because . . . they've raised the ante up to an unbelievable level. . . . If Minnesota is correct, then we are facing in my estimation the biggest environmental catastrophe of human kind. But . . . my gut feeling is that it's an absurd conclusion. . . . this is just baloney" (Cobb 1997).

Three days after the September 30 high-profile press conference, Verlyn and I drove north to the University of Minnesota's Lake Itasca field station, where I'd been invited to give a talk on the deformed frogs for the annual College of Biological Sciences fall weekend for alumni families. When I told my son Steve about this, he said, "Mom, it looks like you've come full circle."

My reentry in 1974 to graduate school for the PhD in zoology had begun at the university's Itasca Field Station, where I took graduate courses in freshwater invertebrates, algae, and aquatic ecology. With my husband out of work, we left Texas and headed to Minnesota where I would start graduate work and our family could be closer to Lon's extended family. At Itasca, we were both without jobs and essentially homeless but for our blue and orange tent pitched in the Field Station's campground. The setting was a far cry from my previous graduate work, ten years earlier, in upper Manhattan at Columbia University.

During the day, my sons, Steve, age six, and Erik, eight, roamed about freely with other kids. They collected butterflies with nets, and after dark they helped pick a variety of insects off a black-lighted sheet strung up in the campground by a friendly high school biology teacher. One day, they unearthed some old bones in a lakeside dirt bank where they'd been playing. The excited pack of kids rushed into the aquatics lab where I was identifying invertebrates, clamoring, "We've found bison bones!" Later the bones were identified, not from an ancient bison, known to roam the area thousands of years earlier, but from a bear, possibly eaten at a lakeside camp one hundred years earlier.

Luckily, late that summer, I landed a part-time position at the University of Minnesota, one that paid better than a teaching assistantship. For some reason I couldn't get a TA in zoology in spite of my high scores on the grad record exams, a cum laude degree from Mount Holyoke and master's in zoology from Columbia. My job at the university's College of Liberal Arts (CLA) paid better and carried a surprising level of responsibility. I participated in reviews of students' academic records and helped make decisions on probation and suspensions, as well as deciding which applicants who had "borderline" backgrounds to admit to CLA. While I had a crash course in CLA's academic policies I was also tackling classes in biometry, microbial ecology, and freshwater pollution. No longer "homeless," we had rented a place near campus and close to a good elementary school for the boys.

As a budding graduate student at Itasca, I sampled zooplankton in the lake at night and logged the impressive, upward distances these almost microscopic animals swam after dark to consume algae near the water's surface. I was enthralled by the diversity of invertebrates like mayflies, caddis flies, decorated leeches, little clams, and tiny rotifers that lived in the lake's various microhabitats. My experiences at the field station, sampling various waters and identifying species in a lab, solidified my commitment to work to protect freshwater organisms.

At that time, in the mid-1970s, researching pollution's effects on aquatic invertebrates (insects, crustaceans, and worms) was largely academic. But today the majority of states use the analysis of aquatic invertebrates to evaluate the condition of their streams and rivers. A high count of the different types of mayflies, caddis flies, and stoneflies, all known to fare poorly in polluted water, is one of several good indicators of clean water. A classic example is the burrowing mayfly, historically abundant in western Lake Erie before pollution caused the loss of oxygen, extinguishing the mayflies in 1953. During the 1990s, after being absent for forty years, mayflies began to recover in Lake Erie, a testimony to pollution abatement programs and other factors. And mayflies appear to be recovering in the Mississippi River in Minnesota after decades of pollution, to the annoyance of people driving across slippery bridges plastered with mayflies.

Soon after my trip to speak at the Itasca Field Station in the fall of 1997, the Health Department's chemical analysis arrived: the water from all three wells had a clean bill of health. Only arsenic was elevated, but it was below the drinking water standard at the time. So why did the well water deform frog embryos?

The well manager from the Health Department speculated by e-mail how this could have happened. Perhaps some new, unidentified chemical was all over the place. More plausibly, he suggested, it could be a low concentration of a common synthetic chemical such as atrazine, widely used for weed control, or its breakdown products. He pointed out that 7 percent of wells tested in Minnesota had low levels of the herbicide atrazine, not typically measured in standard drinking water tests. "Or," he added, "perhaps it's something common to wells, and unrelated to what's in the ponds, that is responsible, like zinc, or arsenic, or low dissolved oxygen."

Interest in the frogs shot up again. By the end of October, the MPCA frog website (including the live Frog Cam) had *40,000* visits from at least thirty countries in only one month. One woman in Holland wrote that she looked at the live frogs every morning when she got up.

Overall, the MPCA and others agreed we had done the right thing by going public with the press conference. People got it, and panic did not ensue. Even the landowner at Dave Hoppe's study site understood. Besides, they had already been drinking bottled water out of fear that something might be amiss in their well water.

At the end of October, the EPA lab in Duluth sent a memo to the state Department of Health and copied MPCA's commissioner. Unbeknownst to us, the Duluth lab had been running its own frog embryo assays with *Xenopus,* a new test for them, with pond water they'd taken surreptitiously from Dave Hoppe's study site. Hearing that EPA staff were taking samples from Hoppe's site, Dorothy, our fieldwork coordinator, was greatly concerned that they might disturb the bottom muds just before her crew arrived to collect mud and water samples.

In the EPA's memo to the Health Department the EPA lab researchers reported that the pond water caused craniofacial and gut abnormalities, consistent with results from our federal partner, the NIEHS. But when they added simple salts to the water, ions like calcium, magnesium, and sodium, the embryos developed normally, they said.

"It appears," EPA wrote, "that these developmental effects are not due to chemical teratogens, but to an ion imbalance/deficiency in the test water." Low concentrations of salts in the water were the likely cause of the deformities, they said.

I had already experienced some uneasiness about the lab researchers. I wished they would share information more openly with us and our partners, not act as competitors, if that's what they were doing. I never heard the results of the toxicity tests they ran on the twelve gallons of water I had delivered to the Duluth lab in late July. Nor did I receive any data from their lab's experiments with the eggs I had collected for them that summer, other than one researcher's casual mention on the phone that he had observed swollen lungs in young frogs and kinked tails in tadpoles.

Bill Souder, who wrote periodically for the *Washington Post,* called to interview me about our test results on the well water. EPA had apparently informed him about their lab results contradicting those of the NIEHS lab. It felt strange to hear about the EPA's research results from a reporter and not directly from EPA staff. Souder said his article, with EPA's comments, was going into the *Post* and we "might not like some parts of it."

That was an understatement.

The article, titled "Colleagues Say Frog Deformity Researchers Leaped Too Soon," ran on the front page of the *Washington Post* (Souder 1997). An EPA associate director at the Duluth lab was quoted as saying, "Now federal scientists are going to look like idiots." He went on, "The NIEHS acted irresponsibly in a rush for headlines. They overlooked some very basic rules for running bioassays. They're not experienced with aquatic species. The Duluth lab is, and the results in Duluth are as clear as you can get." The local paper joined the critique: "Lab Tests on Deformed Frogs Faulty, EPA Charges" (Meersman 1997d, 1997e).

The article referred to EPA's claim that the deformities resulted from low concentrations of various salts in the pond water, not some kind of contaminant. "Results don't mean anything if they aren't interpreted properly," said the Duluth EPA scientist who ran the tests. "In science, spurious correlations happen all the time. It's one of the weakest forms of evidence to support a hypothesis," he said.

"This is incredible!" I exclaimed to Mark. "EPA is discrediting the NIEHS lab results as 'spurious correlations' when they only tested water from *one* site! One! How 'spurious' is that?" I seethed. "And who's *making a rush to headlines* here?"

We'd been blindsided.

NIEHS responded in an editorial, saying the criticism by EPA was "rash" (Lucier 1997). The director of the contract lab working with the NIEHS called me. "Judy, I've analyzed the ions [like calcium] in the pond water, and they're in the acceptable range for running the standardized frog embryo assay," he said. Only potassium was out of range. It was low, but the calcium was okay and that was what was most important, he explained. "There's something else in that pond water that's causing deformities," he said. "It's not just the salt levels."

Ions are electrically charged (positively or negatively) elements or molecules that easily dissolve in water. Table salt, for instance, dissolves into water as sodium and chloride ions, sodium with a positive charge, chloride negative. Calcium chloride and other salts also dissolve as ions. In Minnesota, water in the northeastern part of the state has low ion content (is "soft") because the underlying rock is granite, not prone to erode. In the southern part of the state, more soluble limestone rock prevails and ion concentrations, particularly from dissolved calcium or magnesium carbonate (or sulfate) salts are much higher. These ions make the water "hard," and some homeowners install filters or ion exchange devices to remove or reduce the pipe-clogging, soap-consuming hardness in their water.

It was well known to water quality staff at the MPCA and to toxicologists that the concentration of certain ions, like calcium and carbonate, can modulate the toxicity of some pollutants. Lead, for example, is less toxic in hard water, which is rich in calcium and magnesium carbonate. Lead is more toxic in soft water, which has low ionic strength (low salts). Such realities are built into the state's water quality standards for toxic chemicals. For lead, the standard would be more stringent for soft water because lower concentrations of lead will cause more harm.

Why the EPA Duluth lab manager and its staff scientist took the tack of challenging the results from our federal partner so publicly in the press baffled me.

Then Jim Burkhart called and told me that the analytical lab had passed some of the pond water through an activated charcoal filter, collected the water that ran through the column, and tested it with frog embryos.

"Guess what," said Jim. "There were no deformities, none."

I held my breath.

"The water that passed through the charcoal lost the frog-deforming activity. Whatever it is, it's holding up on the charcoal column," he said, his voice rising a notch.

"It's not the ionic strength," he said firmly.

"This is really good news, Jim," I said. "It means people can use charcoal filters on their tap water to remove frog-deforming chemicals!"

"Yes," he said. In addition, he said, they'd composed clean water with the same, low ionic strength of the pond water and no deformities were caused by the salt concentrations alone. "Clearly there's something else in that pond besides low salts that's causing this," he said.

That our federal partner (NIEHS) had first learned of the EPA's secretive tests through the *Washington Post* was outrageous, especially because the NIEHS had openly briefed the EPA on the results of their lab's recent experiments.

It was a major breach between two environmental agencies, both doing important work on toxic chemicals. The EPA's managers in Washington, DC, had read the *Post* article and were, apparently, infuriated at what their regional lab's staff had done. Later I heard that the Duluth researchers had been ordered not to speak to anyone and were sent to a meeting at EPA's headquarters in DC.

The EPA's actions dismayed me, because I had long respected the work of the Duluth EPA lab on aquatic toxicology, research that helped state agencies like mine set appropriate water quality standards for particular chemicals to protect aquatic life like fish and invertebrates. They'd played a significant role in the MPCA's historic fight against the discharge of asbestos into Lake

Superior from iron mining. And before I landed my job at the MPCA, I'd worked with and admired several of their staff when we tested the effects of pesticides on aquatic invertebrates in ponds.

I expected better.

On the fourth of November, the day after the article ran in the *Washington Post,* EPA staff were back taking samples at Hoppe's study site without letting our field crews know if or when they would be there.

During our communications about work that lay ahead, the well manager at the Minnesota Department of Health had asked what control water our federal partner had used for comparison with the well water they had tested. They hadn't. These well tests were intended only as a preliminary screen of the local ground water to see if it harbored toxicity to frogs. For our next round of well water tests, I suggested we use commercial bottled water for a comparative clean water control. Little did I know what those tests would divulge.

CHAPTER 10

BUREAUCRATIC STRANGULATION

After the reaction to September's press conference and the media coverage cooled down, I hoped work would go more smoothly. But I'd had this thought before. Dorothy took a couple of months off to finish her master's thesis, and, as we expected, she quit her job as fieldwork coordinator when she returned. Without her assistance we faced anew the tensions of trying to manage the frog work and our wetlands projects. Mark and I agreed: this time we couldn't do both.

Dorothy's crews had worked intensely during the summer of 1997. They'd surveyed and confirmed thirty-seven new ponds that had significant numbers of deformed frogs in twenty-nine different counties. They'd shipped numerous samples of frogs and heavy coolers of pond water to several scientists who worked with us. Eighty-three new reports came in from citizens, meaning that since the students' discovery at the Ney Pond in 1995, we had accumulated more than two hundred locations in Minnesota with deformed frogs.

The broader picture was just as grim. Thirty states had submitted records to the federal amphibian center's database on deformed frogs. In Vermont alone, volunteers doing frog surveys found an average of 7 percent deformed frogs in half of the sixty towns they monitored. At the meeting in Prague that summer, I had learned that significant numbers of deformed frogs had been observed in several other countries as well, particularly Russia and Japan.

The epidemic of deformed frogs spread unchecked, while progress toward solving the mystery crept slowly forward. The lab tests by NIEHS showed us that something in the water caused deformities in early-stage embryos of frogs. Their analysis of the water chemistry continued.

Jim Burkhart called me one day with a new result. "Adding thyroid hormone to the pond water prevents the defects from forming," he said, excited.

Jim's lab had identified some suspicious chemicals, like one that inhibited the action of thyroid hormone. But water and sediment samples from several other sites hadn't been worked up yet, much to my disappointment. By then we were all anxious to find answers. I was afraid that Jim's agency would not have the resources after all to tackle the analysis of all the water samples we had shipped from several other ponds. As time passed, I worried that the chemicals in those samples might degrade while being held in their lab's cooler or freezer. Were they exceeding their "holding times," beyond which the results of analysis would not be considered scientifically valid? All that work to collect and ship them. I couldn't think about it.

I began to wonder whether pinpointing the exact cause of the deformities in frogs could be accomplished in only a few years. The causes of more than 60 percent of human birth defects had no explanation. Would frogs be any different? But frog deformities had surged *suddenly* during the 1990s, far more than ever recorded before. That I knew of, such a sharp increase hadn't happened with rates of human birth defects during that time period. Perhaps they surged decades earlier, after the United States dropped atomic bombs on Japan or when US planes doused Vietnam with dioxin-contaminated herbicides.

During the 1990s, something was different in frog habitats in many regions. Frog populations were declining in several areas of the world and malformations were increasing. The widespread and sudden increases in those years raised my hope that explanations would, eventually, emerge. Something had changed, but what?

In the fall of 1997, MPCA staff hydrologists planned to sample the groundwater near additional ponds that had produced deformed frogs. This was a follow-up after our contentious discovery that some homeowners' wells had water that caused deformities in frog embryos. Our hydrologists would install test wells, which they jokingly called "access tubes," to avoid a work slowdown if they applied for well-drilling permits from the Health Department. Would groundwater from other locations than the handful of wells tested that summer also be toxic to frogs? Might contaminated groundwater be seeping into the shallows at the edges of frog ponds?

As winter approached, the ground was starting to harden, and our hydrologists needed to get out to the sites right away. With Dorothy gone, I attended their meeting in case they needed background information. I expected to listen in on a thoughtful discussion about which ponds to drill and sample, or to offer advice on shipping groundwater samples to the federal lab for the frog embryo tests. By then, I had intentionally reduced my efforts on the frogs so I could wrap things up before I retired, whenever that would be.

We gathered in a gray, windowless conference room and sat around wooden tables. I greeted a couple of the MPCA hydrologists I knew and had previously worked with. They were dressed casually, as if ready to get out in the field. Another hydrologist, sporting a dressier shirt, grinned and stood up to run the meeting. More a desk worker than a field hydrologist, he hadn't had anything to do with the frog investigation, and I knew little about him.

Forty-five minutes into the meeting, without a word uttered about plans for the groundwater testing, I felt a growing sense of uneasiness. The meeting leader droned on. One of the field hydrologists leaned his head on his hand. I sagged. *How long will this take?*

"We need to develop individual health and safety plans for everyone who goes out to deformed frog ponds," the speaker said. "For that we need to get a safety expert, who will have to interview staff in person before he can write up each individual's safety plan. Also, we need to find a secure place to lock up all the pertinent files on the frog work, a place where we can lay out our maps." I clutched. *Lock up the files?* Would our ready access to our site files be crippled, guarded by a bureaucratic gatekeeper? I suspected this staffer was trying to be careful and do things the right way, in his view, and I don't think he intended to cripple the frog investigation.

"We need to set up a regular, weekly meeting time," he continued, "and we need to meet frequently with the other state agencies, so we don't step on anyone's toes or cause any brush fires." I shook my head in disbelief. We had *no* time for this. None!

My fears intensified as I realized this fellow wanted to take control of the frog investigation, not just help the hydrologists make plans for their imminent field sampling. Where had he gotten this idea? Had someone directed him to take charge? I began to imagine a dark hole forming in the middle of the conference table, a chasm about to suck in the fall's groundwater sampling efforts, if not more.

"We should start a project-planning exercise," the staffer continued, as he passed around thick copies of a handout. I glanced at it. My anxiety grew. The material reeked of the MPCA managers' business-oriented deliberations of late. “Big-picture brainstorming process” caught my eye: “We should not limit ourselves to linear and conservative thinking. Instead we will loop back and cycle through the process,” I read. I almost laughed out loud over the irony. The linear thinking that prevailed among our leaders and at this meeting already impeded doing good science at the agency, at least on the frogs. I seriously doubted that this paper exercise could change that.

The rigidly defined “planning process,” perhaps well intentioned, was utterly divorced from the ongoing realities of the frog investigation. The meeting leader had no idea of the many partnerships and different kinds of analysis that were already underway, of our flexible and evolving, *nonlinear* work plans that might not fit into some bureaucrat’s desk-designed diagrams. Worse, this meeting leader proposed a *new* bureaucratic reality, one that could kill the science of the frog investigation and prevent any real work from getting done.

I watched my imaginary hole widen and fill the room. Could it swallow the entire investigation? I foresaw the future: *death of the frog investigation by bureaucratic strangulation.*

I remembered what one scientist had said at the conference in Prague. Concerned about the future for amphibians, he had quoted Einstein as saying that serious problems cannot be solved at the level of thinking that created them. I’d been quite humbled by the complexities of the frog investigation and wasn’t sure that any of us had the level of thinking we needed to solve the deformities. But this was not it.

Whether the staffer’s attempt to control and stifle the frog work was mindless bureaucracy or intentional didn’t matter. Either way, something had to change, and fast.

During this time I was having dreams about rescuing babies in trouble. In one I cleaned and dressed a toddler’s wound. It got back on its feet. In another, I rescued a baby about to smother in bed covers that had spilled over the side of a bed. I carefully placed it in the center of the bed so it could breathe. These dreams worsened over several weeks as the perils for the frog investigation increased at work. I dreamed I discovered a baby, with its mouth open under water in a bathtub. I pulled the baby out, laid it over the tub’s edge, and expelled water from its lungs. When its legs moved, I felt a great sense of relief. In the final dream, I saved a baby from electrocution.

All the while, MPCA management moved inexorably toward its much-touted “Goal 21” reorganization, scheduled to take place in the summer of 1998. Major divisions, like Water Quality, Air Quality, and Hazardous Waste, would be dissolved. The new structure would take on a more regional focus. Months of meetings had consumed the managers’ time and agency’s resources, and staff worried deeply that the MPCA was losing its environmental and regulatory focus.

One day, an all-staff memo defining "Goal 21 Vocabulary" appeared on my computer. "Mark, have you seen this?" I said, frowning. "It defines things like 'situational alliances,' 'hard-wired teams,' 'causal loop diagrams,' and 'behavior charts.' Check it out." I squirmed uneasily in my chair and added, "Sounds like some kind of business-speak."

We searched the list of twenty-nine Goal 21 definitions and terms to find the word "environment."

"Here it is!" Mark exclaimed. He found "environment" couched in one of the four goals for Goal 21: an aim of "cooperating with customers to create a broad plan to protect Minnesota's environment." Mark grimaced.

"At least it's not saying we have to make a *profit* by protecting the environment." He looked at me and grinned.

In December of 1997 my position was finally converted to permanent (or classified) status. By then I had worked at the MPCA for eight years with no job security. I'd been kidding Verlyn that I'd retire never having had a secure job in my entire life. Oddly, I felt no surge of joy when it proved otherwise. Through many risky times at the agency I had yearned for a permanent assignment, to have my work accepted. It could have been an honor, because at that time only a few women held the position of Research Scientist III. Then I understood why a woman biology professor I knew at St. Olaf College seemed almost bitter when, in 1988, she became the first female full professor in the natural sciences. It had taken far too long. Also, any celebration of my new status was tempered by our director's threat earlier that once Mark and I were placed in classified positions, we might be forced to do work we wouldn't care about. Would this promotion mean I'd be pulled away from the work I loved and valued?

We rolled into 1998, the year I suspected might be the last field season for the frog investigation. By then, both the frogs and the MPCA were in trouble: frogs because the deformities were so widespread and unsolved, and the MPCA because it had dragged its feet on other major pollution issues. Representative Munger and the environmental groups were upset. The agency had been slow to regulate feedlot pollution and had done little to prevent contamination that was traveling insidiously underground from a huge oil and gas refinery toward the Mississippi River. With the MPCA's pro-business orientation, such laxity wasn't surprising. The agency was roundly criticized in the local press: "State Watchdog Lags in Policing Water Pollution" (Rigert and Ison 1998); "State's Watchdog Losing Its Bark, as Well as Its Bite" (Grow 1998).

At least this time around the agency said it was willing to receive more funding for the frogs from the legislature. Munger authored a new bill for $500,000 for the MPCA "to accelerate the research on deformed amphibians" in 1998. I was asked to testify about the frog investigation at the MPCA's budget hearing in the legislature.

I rushed over to the State Office Building (sometimes called "the SOB") for the hearing, and ran down the stairs to the large, wood-paneled hearing room. People waiting to testify and other observers sat in rows of seats to each side of the central tables set with name plaques for the legislators.

As I entered the hearing room, Representative Munger, by then eighty-seven years old, asked me to sit beside him at the presenter's table that faced the legislators, a great honor for me. I doubt Munger knew how much his affirmations and support of my work had sustained me. By then, he had become sharply critical of the MPCA's lethargy in regulating polluters. He had recently held a hearing to allow several environmental organizations a chance to air their grievances against the MPCA. The hearing was so tough that our agency's assistant commissioner reportedly broke down in tears.

At the budget hearing, the committee chair, a legislator known for his negative attitude toward the MPCA, began with his usual jabs against the agency. First to speak was an MPCA manager, who presented the overall budget. I looked around the hearing room: one representative was chewing gum open-mouthed; a couple of legislators were paying no attention but were whispering to each other, perhaps about some other issue. One was hunkered over, eating pizza for lunch.

I stood up to speak about the frog work. The chair began on a pugnacious note: "First you thought it was in the water. Now you say it might be ultraviolet light. You really don't know what's doing this, do you?" he jeered.

I said we were sure that the cause was something in the pond water and told him that chemicals in the water were being identified. I tried to explain to him how scientists work with different ideas about what could have caused the deformed frogs, how they gather data and evaluate it. We expected to narrow the field of possibilities and would likely rule out both ultraviolet light and parasites after the coming year's work, I said.

Only one lawmaker actually seemed interested in the science and asked me some astute questions about farm chemicals and their breakdown products. What were we doing about this, the legislator asked. I replied that we had an expert scientist from the US Geological Survey measuring pesticides and their breakdown products in frog ponds.

When I described our need to know which pesticides were used around the ponds that had deformed frogs and our difficulties getting that information, the chair grumbled. He didn't want us looking into how citizens use their land as a cause of deformed frogs. Doing that intrudes on their private property rights, he said.

My heart sank when he announced that any new funding the legislature might provide for frog work at the MPCA *could not be used to pay agency employees to work on the frogs.* How could we work in the field without hiring (and paying) seasonal staff? Mark and I were still covered by our EPA wetlands grants, but new money was necessary to staff the field work for the frog investigation.

Just when I felt worn down by him, he caught me off guard by saying, "Dr. Helgen has given me a number of good reports. At least MPCA has *one* good employee." After the hearing, the committee chair passed by me and said, "You're fun and you're good." I couldn't take this as a complement. Too often I'd seen him torment people in legislative hearings, like an animal batting around its prey. Representative Munger paused to praise me and shake my hand—and apologize for the committee chair's attitudes.

Jim Burkhart called me late that fall. Speaking in low tones, as if he didn't want anyone else to hear, he said, "Judy, those two samples of bottled water you sent to us to use as control water to compare with the pond water? *That* water is causing deformities in the lab tadpoles!"

I couldn't believe Jim's words. "Even bottled water?" I blurted. I told our hydrologist and Mark, who immediately called the bottled water company for copies of tests they'd done on their water. Their report came back showing the water tested clean, at least based on standard drinking water analysis. Was it some chemical that had not been analyzed or something leaching from the plastic jugs that deformed frog embryos in the lab?

The public erroneously believes that bottled water is as safe if not cleaner than regular tap water, that it is routinely tested. But studies show otherwise: in the late 1990s the Natural Resources Defense Council (Olson 1999) analyzed more than one hundred brands of bottled water and found contaminants that were in violation of drinking water standards in 22 percent of samples. Only 45 percent of bottles they tested had no contaminants or causes for concern. Contaminants included metals (arsenic, for one), perchlorate (a chemical used in rocket fuel), and intestinal bacteria and fungi. Some bottled waters are actually nothing but tap water, repackaged. But that I knew of, no one had previously tested bottled water to see if it caused deformities in lab animals, such as frogs.

I found out from Mark that the fellow who'd run the tedious and disastrous planning meeting for the groundwater sampling had been told by our field hydrologist about these latest results.

As I rode the elevator down on my way home, he got on.

"The results that the bottled water causes deformed frogs our hydrologist just told you about? We need to keep the lid on that until we find out exactly why that happened," I said. It was just a phone call. There might be some explanation. "Please don't tell anyone," I said.

As we walked off the elevator, he said with a sly smile, "Well, I have told one person—my section manager." I couldn't believe it. I rushed to a pay phone and called Mark upstairs. "Oh good grief," Mark said. "This will pass among the managers, who will likely tell the Health Department before we've even *seen* the results!" Which is what happened, sending our relations with the other state agencies further downhill.

After this episode, I had to do something. I talked with a manager and the field hydrologists about the role of this staffer, who apparently wasn't aware of the political volatility that surrounded the work. I raised the idea of moving him off the project. The frog investigation was far too sensitive to let this kind of leak happen before we even had a chance to examine and interpret Jim's data. They agreed.

This event added fuel to my growing suspicion that we could not do science in the climate we had at the MPCA. Even preliminary results reported to me in a phone call traveled, like wildfire, within a matter of hours to another state agency.

It also meant that I would have to stay actively engaged in the frog work when I needed time to focus fully on the development of the aquatic invertebrate index for wetlands before I retired. Almost reluctantly, I committed myself to form plans for the frog investigation with several cooperating scientists and agencies for the 1998 field season. In the few years I had before retiring from the MPCA, I would be the lead writer for the EPA on a "module" about the use of aquatic invertebrates to monitor wetlands quality. That document synthesized the ongoing work in several states (Helgen 2002a; US EPA *Wetland Bioassessment Publications*). I would refine and complete the guide book that our citizen volunteers used for sampling aquatic invertebrates from wetlands, identifying them, and calculating a score to rate each wetland's quality (Helgen 2002b).

As we faced 1998, I knew I couldn't abandon the frogs.

To make plans for 1998, we had to keep our focus on the frogs and not succumb to the political winds that worked against the investigation. Our best bet was to continue looking for causes in the habitats where the deformed frogs lived. I proposed to our research team that we follow the framework referred to as "eco-epidemiology" and used by the scientists who pinpointed certain transformer oils (PCBs) as the cause of deformities in herring gulls and other fish-eating birds around the Great Lakes (Fox 1991). In brief, some of the criteria were:

1. The putative cause of deformities is likely to be present at key times in development;
2. the causal agent shows a dose-response relationship with the deformities (i.e., more of the agent causes more deformities);
3. in the lab, the causal agent triggers deformities similar to those seen in the wild.

I assembled a detailed work plan with our collaborators, which included the USGS, the National Wildlife Health Center Lab in Madison, the NIEHS in North Carolina, some academic researchers, and a few technical contract labs. The plan included tables listing which types of sample would be collected and delivered to the different labs.

Pond water, bottom mud, and groundwater would be delivered to several labs for various kinds of chemical analysis and the frog embryo test. Four different researchers would diagnose diseases, do X-ray imaging of frog skeletons, analyze tissues for pollutants, and look for the presence of parasites. An MPCA crew would carry out the on-the-ground fieldwork, collecting frogs and pond samples and shipping them to the researchers' labs. Staff helped develop a "quality assurance" plan for the work.

Scientific studies for government research are guided by protocols for what's called "quality assurance," especially for environmental analysis that might be used in a regulatory action or lawsuit. In an investigation of a suspected chemical released by a polluter, for instance, plans are reviewed by agency staff with expertise in quality assurance. Guidelines must be followed for the number of replicate samples taken, for the cleaning of sample bottles and equipment, for labeling, icing, and delivery times to the labs that carry out the chemical analysis. There are designated "holding times" for how long a sample can be stored before it is analyzed, because some chemicals degrade over time or react with other chemicals in the water or sediment. Labs that perform the analysis must use EPA-certified methods and analyze control samples, like clean water and "blanks" to confirm that the sample jar itself is not chemically contaminated.

In a legal or regulatory action, a "chain of custody" for each sample must be established to verify that no one could have adulterated the water or sediment between the time it was collected at a field site and analyzed in a laboratory. In some cases, "blind coding" is required. Samples are labeled with codes so the testing lab has no clue as to the location or origin of the sample. This is to avoid any possible bias in the results or the interpretation of them by the testing lab staff.

Our proposed plans were reviewed by three expert scientists who were not directly involved in the frog work. Tim Kubiak, a contaminants specialist with the US Fish and Wildlife Service, wrote that we were on the right track. He discounted parasites as a probable cause of the deformities and recommended that we put most of our effort toward identifying which chemicals the frogs might likely encounter.

He called me to ask which pesticides had been used near the frog ponds. "We can't get that information," I said. In an earlier meeting, a manager in the state's Department of Agriculture had told Mark and me that this data, which they had but wouldn't share with us, "was proprietary, or confidential information," as he shuffled his feet under the table.

"You're kidding me," Kubiak said, incredulous. "That should be public information." He offered to go behind the scenes to see what he could find out. But I doubted he could pry loose the closed doors of the Agriculture Department.

A Canadian scientist, Mike Gilbertson, who had participated in the earlier investigation of the causes of Great Lakes bird deformities, also reviewed the work plan. The plans had "a balance between field observations and surveys, pathology and analytical work," he wrote. In a later phone call he described the frog research as "forensic" science. He told me his work on the bird deformities had also faced political and bureaucratic roadblocks. "I think you are handling this fine," he said, encouraging me to stick with it. My grandchildren will be proud of me, he commented.

With the critiques in hand, I sent copies of the final, revised, and peer-reviewed work plan to our sister agencies—the state's Departments of Health, Agriculture, and Natural Resources.

Now we could move forward. Or so I thought.

One day I stopped by the office of my new supervisor, Marge. She had recently been assigned to oversee staff working on the frogs but had little background about the history of the frog investigation, nor the political issues associated with it. Marge had just suffered through another grueling meeting about the frog investigation with high-level representatives from other state agencies, meetings I was not asked to attend. I knew they'd be reviewing our twenty-page plan.

"How did the meeting go with the interagency group? Did they like our work plan?" I asked.

"Judy, I don't even know how to tell you this, but the sister agencies want the MPCA to *stop* the current work plan and shut down all work on the frogs! They're asking us to step back and take some time to figure things out." For one thing, she said, the agencies wanted a broader-scale, *randomized* survey of ponds, not the targeted approach we had planned, studying several pairs of sites that had either deformed or normal frogs. "What's more, before any work is resumed, the agencies want to form *three* interagency committees!" She sighed, her exasperation evident. "They want one committee to manage the frog investigation, one to be a technical advisory committee, and a third to serve as overall advisory group." I rolled my eyes, barely able to absorb her words.

I stood in the doorway of Marge's office, staring in disbelief. Was this it? The end? The ultimate bureaucratic death blow? If we stopped work at this point to "figure things out"—whatever that meant—we'd lose the upcoming field season, perhaps the entire investigation. Our partners stood ready to go, and soon.

"You've got to be kidding," was all I could get out, thinking of our complex and detailed work plan with several researchers and federal agencies.

I remembered how a staff person had worried that we'd "step on toes or create brush fires" with the other agencies. I knew we had annoyed the Agriculture Department when we asked for their lab's sampling protocols and for their closeted information on pesticide use. And when the federal lab discovered that private well water caused early stage deformities in frog embryos in lab tests, a manager in the Health Department became furious, even though the authority over private wells was under the MPCA's purview, not theirs. Health's response had been to deride the frog embryo lab test. "That test doesn't mean anything," said one of their staff during a meeting, while others chuckled. I understood that the Health Department would bear the brunt of worries about drinking water and threats to public health. But why block an investigation into the causes of frog deformities?

Now we had a new and unexpected nightmare—that the other state agencies would take over and collectively strangle the frog investigation.

"We can't stop now," I said to Marge in a low voice. "If we do, we lose the critical time of egg laying." I could see her genuine frustration. The other agencies had been sending high-level managers to the interagency frog meetings. But our division director had sent Marge, who knew little about the politics and science of the frog problem. Without support from the MPCA's leadership, the frog investigation was now trapped in a mortal struggle among the state agencies.

I stood in Marge's office, speechless. Two state agencies were telling us to stop the investigation and MPCA leaders were not backing us up.

Overwhelmed, I felt I would have to quit, if not my job, at least the frog work. Recently, Verlyn had expressed concern about my level of job stress, my lack of sleep, and my frequent need to work late at night and on weekends (even on Easter Sunday).

I was trying to find some sort of balance in my life. I had scheduled a short stay in a hermitage cabin at a retreat center run by Franciscan sisters at a farm north of the Twin Cities. During that retreat and on other occasions over supper with Verlyn or with my close female friends, I brainstormed job alternatives. Could I leave the MPCA and consult? On what? Create the course I wanted to teach on pollution biology? Before coming to the MPCA I'd taught a popular course on the biology of women. Perhaps I could update that and teach it again. Could I write a book about vernal pools or about the frog investigation? Work on environmental issues for a nonprofit organization? I thought of my mother, who learned to paint beautiful landscapes later in life after her early education in the sciences and a few years teaching biology became eclipsed when she had a family. I had taken and loved classes in figure drawing years before. What could I do now that would be more creative than my job?

My older son, Erik, was reframing his life at age thirty-two. After working various jobs for years after college he decided to go into architecture and headed East to grad school. Before he pulled out in his overpacked, rusted Honda Civic, he and I battled to get his struggling cat into its cage for the trip. He could launch himself into a new life, could I?

Contemplating quitting made me feel weak and undedicated, as if I lacked the level of commitment I knew respected environmental leaders had—like Representative Munger, who consistently fought to protect Minnesota's waters and habitats, or scientists who'd worked successfully under political pressure to show that PCBs or selenium caused deformities in wildlife. What kept them going?

I bumped into a staff person, a gentle former pastor, at the copying machine. "You know a lot is riding on this project," he said. "This has got to be the biggest thing this agency has ever been involved in. We've got to have it right." Shamefaced, I crept back to my desk, not able to tell him I was on the verge of quitting. I felt split down the middle. My heart was with the frogs, with my hard-working colleagues, and with the kids. Could I hold on until after the fall elections, when a newly elected governor might bring the MPCA back on track and shift the political leanings of upper management? Could I last that long?

One of my father's mantras, sometimes expressed jokingly in silly situations, sometimes seriously, had been, "Strive on regardless!" His caring voice sounded in my head. *Strive on regardless, Judy.* At times of struggle, he had always supported me, especially during the difficult period after my first husband's stroke, when I had a four-month-old and a toddler to care for. Dad phoned, wrote letters, gave important advice on finding good medical care, and generally helped me keep going. He sustained my ability to make tough decisions, to plan, and to cope with the immediate and longer term consequences of my husband's desperate situation and serious illness, with our future.

If only I could consult with him now.

What would happen to the seasonal workers we were hiring if I left my job? And the frogs, the most vulnerable of all: would my leaving bring down the already teetering investigation? I thought about all the people who were worried about the frogs and wanted answers. I knew I *had* to stay, at least to shepherd our new work plan through the maze of competing interests. Feeling powerless over agency politics, I vowed to do whatever I could to keep the frog investigation alive.

At the last minute, the MPCA reversed course, again. The assistant commissioner finally stood up for the frog investigation and brought the other state agencies around. The work plan was given the go-ahead, just in time for spring. By now I was deeply skeptical about the repeated roller coaster of changes at the MPCA. Absent political will and consistently supportive leadership, carrying on a contentious environmental investigation there seemed almost impossible.

Barely in time for the spring field season, we hired a new fieldwork coordinator, Sue Kersten (now Vanden Langenberg), a positive, organized young woman. She had researched the effects of contaminants on frog reproduction for her master's degree in Wisconsin, and she was savvy about amphibians in both field and lab settings. Sue represented hope for the investigation's future.

Sue took charge and cheerfully geared up to collect early-stage tadpoles for three researchers who would look for evidence of disease, parasites, genetic damage, and key types of deformities, work she was familiar with. Luckily, we rehired Jeff Canfield, who knew the field sites well, having worked with us the summer before. Jeff was quiet but solidly competent, and he knew his way around the frog database. He and Sue made a terrific team.

But their positions were vulnerable and temporary, dependent on the money the MPCA had publicly promised for the frog work. Somehow, half of that, around $150,000, had mysteriously disappeared somewhere into other MPCA accounts, Mark told me one day. We suspected it had been funneled into the pockets of outside consultants the agency kept hiring to assist its ongoing reorganization, the latest attempt called "the course correction."

I mentioned to Sue early on that we'd just had a relatively calm day. She put her head in her hands and joked, "If this day was calm, what are the *others* like?" Knowing the pressures we all faced, Sue and Jeff tried to lighten things up occasionally. One morning I picked up my phone and was greeted by a cacophony of rising amphibian trills, courtesy of Sue with her cell phone pointed to an orgy of breeding toads. "Just to cheer you up!" she said brightly. She and Jeff were traveling to the study sites to hunt for frog eggs and locate the gelatinous egg masses precisely by using geographic positioning units. From the Ney Pond, she called, all excited. "We're seeing at least a hundred egg masses, I wish you could be here!" Then there was the croaking plastic frog Sue hid under my desk . . .

One day Sue, Jeff, and our hydrologist, their faces suppressing grins, strolled into our work area. "We've solved the frog problem," one of them said. "We've discovered something the deformed frog ponds have in common! Budweiser cans! We've found them at each site!" We roared with laughter. At that time, frogs had starring, if bizarre, roles in Budweiser's publicity campaigns. For reasons known only to beer marketing people, frogs croaked in Budweiser's ads on TV and were pictured leaping across the side panels of their beer delivery trucks. A cartoon by Steve Sack in the Minneapolis paper depicted three goofy-looking deformed frogs with "Bud-wei-ser" written across the top. Doug Wilcox, a prominent wetlands scientist, sent me his home-drawn cartoon: a Budweiser truck in the background, normal frogs watching a extra-legged frog across a pond. One says, "I heard his parents met on a beer truck." We posted both cartoons at the entrance to our work area.

After one trip to Dave Hoppe's intensive study site, Sue and Jeff rushed breathless into our work area. "Something's wrong with the egg masses at Hoppe's site," Sue said, explaining that the eggs looked like black specks in the jelly. Jeff chimed in that minnows were floating dead, as they had last year.

"Hoppe's afraid the leopard frogs are in real trouble," Sue said. He worried that the few frogs he'd seen weren't mating, she added; his automatic tape recorders were capturing very few frog calls at night. Alarmed by this ominous report I wondered if we were headed for an amphibian silent spring. I thought about chemicals that could interfere with the development of the male frog's larynx, which is governed by both thyroid hormone and male hormones in early development. If males couldn't call, females could not find them to mate in the night.

Eventually, management permitted me to meet with staff from the other state agencies and explain in person what we were doing and why. With Sue Kersten becoming more experienced at coordinating the frog work, I reduced my role in the investigation. After working six years on the deformed frogs on top of my regular job, I could finally focus on wetlands.

By then, most people had acknowledged the serious nature and the potential, wider significance of the deformities that afflicted so many frogs. We had no prime suspect—nothing like the metal selenium that leached from soils into agricultural drainage systems in California, causing severe deformities in wildlife. Scientists were grappling with the possibility that multiple causes of deformities were in play, and were asking how various factors such as chemicals, UV radiation, diseases, parasites, and even predation might interact. Did depressed immune systems play a role? Chemicals that disrupt normal hormonal systems? What was happening might not be resolved by lab tests of single chemicals on aquatic species.

The problem was real. At issue was the need for more research on all fronts, by both government and academic scientists. We faced the reality that determining the causes of deformities in frogs might take decades.

Some progress was being made, however. Our federal partner (NIEHS) had evidence that something in a pond's water and its mud could cause early stage deformities in lab-reared tadpoles, and this suggested that chemicals were the prime suspects. Moreover, most of the deformities triggered by pond water in the lab could be prevented by adding thyroid hormone. So something interfering with normal thyroid hormone activity could be a factor, if not the cause. Filtering pond water that caused malformations through activated charcoal prevented abnormalities from developing, which meant the causative agent could have adhered to the charcoal particles.

The federal lab had identified several chemicals in the water collected from one pond by the MPCA field crews (Fort et al. 1999a, 1999b). Four of these caused the greatest number of defects when tested individually in the lab with frog embryos: a crop fungicide (Maneb), an insecticide (permethrin), the metal nickel, and propylthiourea, a chemical that inhibits thyroid hormone. A lot more work like this was needed. What about the water in the other ponds?

We knew that pesticides got into frog ponds from farm runoff and precipitation. During the frog investigation, scientists from the USGS detected pesticides in rainfall they had collected in devices installed alongside two of our study sites. They also found several pesticides in water that flowed through the four underground agricultural drainage tiles that conveyed farm runoff into the Ney Pond (Jones et al. 1999). Atrazine, cyanazine, metolachlor, and ten pesticide breakdown products were found, but the concentrations of each were less than three parts per billion. Were these the culprits? Did low levels of such pesticides or their breakdown products harm tadpoles? Could mixtures of them act together? Who knew? We were stymied by the limited research on pesticides (or other chemicals for that matter) and frog development.

Years later, in one book (Collins and Crump 2009, citing Souder 2000), the Ney Pond was mistakenly described as having "no obvious disturbances . . . no incriminating pipe discharging an unknown fluid, and no hazardous dump in sight." In truth the Ney Pond had four pipes discharging agricultural runoff into it; it had been excavated and created in the late 1980s; it was surrounded by cropland. A University of Minnesota lab detected, inexplicably, radioactive iodine in pore water in one of the small, created islands in the Ney Pond. The pond had a history.

Could a change in chemical usage in the early 1990s have explained the epidemic in frog deformities that surged during the 1990s? One widely used herbicide, cyanazine (Bladex), could have caused malformations because it had a chemical structure that was known to cause skeletal abnormalities in vertebrates (an organic nitrile). From 1990 to 1993 cyanazine was the fourth most heavily used herbicide in the country. USGS had measured cyanazine in rainfall in the early 1990s in other areas in Minnesota and in the groundwater in many sites in Iowa (Capel et al. 1998).

The USGS also detected cyanazine in the agricultural runoff that discharged into the Ney Pond and in rainfall collected at Dave Hoppe's site in north-central Minnesota, but not in the pond water there (Jones et al. 1999). Eventually, cyanazine use was halted by the EPA because of reports of a variety of birth anomalies, including missing eyes, and some results that showed increased mammary cancers in certain lab animals. Production stopped in 1999, but the EPA allowed use of existing stocks through 2002.

I could only speculate that cyanazine caused the frog malformations, but this is one example that fit the idea that a pesticide or some chemical additive that was widely used during the 1990s was responsible. If such a chemical was subsequently withdrawn from use, as was cyanazine, then one could hope that frogs might once again develop normally.

CHAPTER 11

THE WRECKING BALL

The MPCA's new reorganization plan, optimistically called "Goal 21" for the new century, swung into action with a massive and work-stopping internal move in 1999. During this time, the agency received a bomb threat, preceded earlier that summer by an anonymous message from someone who had threatened to "start shooting employees one by one," should the MPCA go ahead with its plans to fine two communities for violating pollution regulations. This made staff uneasy and led to security systems at the building's entrances.

The costs of adding and moving staff to regional offices located in cities away from the metropolitan area and the fees for the expensive consultants who'd helped plan the reorganization burned a hole in the agency's budget. Within a year the MPCA was at least $6 million over budget and accounting systems were in disarray. All nonpermanent staff were asked to leave, including the summer seasonal staff who supported biological monitoring of streams and those who would help Sue and Jeff survey the frogs. Without summer assistants, fieldwork would grind to a standstill.

One day, a staff person from accounting came by to tell Mark and me that our EPA wetlands grant was overspent because it was paying the salary for an employee who worked in another division.

"You're kidding," I said.

"Well, there's more that you guys should know," she said. "The agency has already spent down your *next* EPA wetlands grant."

Mark's jaw dropped in disbelief. "But that grant hasn't even kicked in yet—that's impossible!" he said, his face pale.

"You're going to have to fight to keep your wetlands money every step of the way," she said.

Our newly elected governor, former wrestler Jesse Ventura, had made it clear that he wanted to cut state government. He openly said the MPCA should be "scrapped" and regulations left up to the federal government. His proposed budget for the MPCA was severe: the agency would have to cut seventy of the seven hundred or so positions. Later, during the strike by state professional employees in 2001, Ventura came out to talk to MPCA and Health Department staff who were picketing outside the governor's mansion. A colleague who was there described the scene, with the governor leaning on the wrought-iron fence, his cigar in hand. He knew a lot of people who would like the strikers' jobs, he said. "But can they do hydrogeology?" one scientist asked. At that Ventura had walked away.

To accomplish Ventura's cuts, the MPCA's upper-level management developed a list of agency programs to eliminate and sent it out for all to see.

I looked at the program elimination list with horror. "Mark, they're cutting out the wetlands positions!" I said.

In addition to the proposed elimination of our jobs for the wetland bioindicator work, the agency would also drop a federally required program that served to protect wetlands. Under the Clean Water Act, the MPCA would no longer review potential harms to water quality from activities like dredging and filling in wetlands, dredging rivers, or building dams. The program (called 401 Certifications) is required in all states by the EPA and has been considered the primary means for regulating activities that affect wetlands, an important way to try to protect them. Texas and Alaska had unsuccessfully attempted to eliminate their 401 programs. Only Minnesota would actually drop its wetlands permitting responsibilities. Wetlands protections were falling away.

The Clean Water Act (CWA) water quality program that oversees alterations to wetlands is administered by the US Army Corps of Engineers because the Corps had decades of experience in regulating dredge and fill activities that affected navigation and shipping. When the Corps, not the new EPA, was assigned CWA permitting responsibilities over wetland-damaging activities after the passage of the Clean Water Act in 1972, many saw this as a giveaway to developers and to oil and gas interests (Vileisis 1997; Pittman and

Waite 2009). At first the Corps tried to narrow the scope of its responsibility over wetlands activities to those wetlands that affected coastal and inland "navigable waters." But in 1975 a court action that resulted from a lawsuit filed by environmental groups (*NRDC vs. Callaway*) asserted that "navigable waters" in the Clean Water Act includes all waters of the United States. In 1987, the Corps published its manual for determining which waters can be defined legally as wetlands (US Army Corps 1987).

In 1990 the Corps agreed to adopt the EPA's guidelines for working through wetlands permits, a process called "sequencing." First, any impact to a wetland should be avoided; second, the impact should be minimized; and finally, if impact is unavoidable, then the damage to one wetland must be "mitigated" by creating or restoring a wetland elsewhere. Often developers skip to this final step. In some states they can purchase "credits" for wetland acreage deposited in a questionable wetlands mitigation "bank," which lists wetlands that supposedly are already restored or created. The EPA has veto power over all projects permitted by the Corps and states, although it rarely uses it.

Considered valuable public resources, at least on paper, wetlands were not to be altered or destroyed. The Corps would review permits from developers, farmers, and others who wanted to alter wetlands for their projects. State pollution agencies (such as the MPCA) would review each permit to determine whether the wetland-altering project would harm the state's water quality, then sign off on it.

How protective this Clean Water Act program really is for wetlands remains to be seen. Does the permitting process at least slow the destruction of wetlands? Not in Florida: see Pittman and Waite 2009.

In 2001, citing "budget constraints," the MPCA began waiving most of the Clean Water Act permits that came to the agency from the Army Corps of Engineers. MPCA staff would pass them on without any review of the impacts of a proposed alteration of wetlands. Not until 2007 did the MPCA restart this program and resume its water quality review of wetlands-altering projects that required Clean Water Act permits. Of course, such reviews remain unpopular with commercial interests and landowners.

At the MPCA, things had spun badly from 1998 and after. Management had proposed that the positions for our wetlands monitoring work and the staff that reviewed wetlands permits be eliminated. In addition, the fledgling biological monitoring program was in trouble. Its budget had been raided, and its framework was in jeopardy. A few streams that had been identified as very polluted (based on poor scores for their fish communities) had been placed on the EPA's required national list of polluted waters. A breakthrough for the biological work, we thought. But unexpectedly the MPCA reversed itself and *removed* these streams from the EPA's list, claiming the biological scoring systems were never intended to be used for listing impaired waters in Minnesota. An internal battle ensued, and some time later the agency did restore these poor-quality streams to the EPA's impaired waters list.

Mark and I had been moved back into the biological monitoring group, which was led by a supervisor, Greg Gross, who had experience in water quality standards and was supportive of using biological indicators. I asked him why the MPCA had taken the streams off the polluted waters list. He explained that local people had said they wanted greater protection for fishing lakes than they did for those particular streams. Was the MPCA shedding its responsibility under the federal Clean Water Act? Was the MPCA going to let local people decide which pollution is acceptable and where? Management touted a shift to local control as a good idea.

During this time, I talked with Chris Yoder, a scientist who had led Ohio's successful program to adopt and use biological scoring systems to rate the water quality of Ohio's streams. I had great admiration for Yoder's pioneering work in Ohio, a state that continues to assess surface water quality by analyzing the fish communities and calculating a score for the Index of Biological Integrity, or IBI (Ohio EPA 2010).

The fish IBI is composed of several "metrics," or measures of the fish community. For example, lower metric scores are assigned when there's a high percentage of species that tolerate pollution, or when a certain number of fish exhibit deformities or lesions. Higher metric scores are given for indicators of better water quality, such as a fish community that has a greater percentage of carnivorous fish or a higher proportion of insectivores (insect-eating fish). The IBI score is the sum total of scores for eight or more different metrics. Today the MPCA also uses the fish IBI to monitor and rate the water quality of its streams. (For basic information on the IBI, see Karr and Chu 1999.)

For years Yoder had actively promoted the idea that healthy fish communities indicated better water quality in streams. He demonstrated how the IBI could be used as a money-saving tool because it better targeted stream sites that had the most impoverished fish assemblages as ones needing more pollution abatement. Subsequent improvements in IBI scores showed where pollution controls made a difference.

I told Yoder about my agency's staff cuts, about its desire to avoid the "old command and control" approach to regulating polluters, about the raids on our budgets. He paused, then said, "Bit by bit, protections for the environment are being dismantled all over the country. It's like taking the flesh off the bones of environmental protection so gradually that people don't even notice it, until one day only a skeleton remains," he said.

These shifts in fundamental attitudes about regulation at the MPCA fit in with newly elected President George W. Bush's pronouncement in 2001. Soon after entering the White House, Bush instituted a freeze on new government regulations. He called for major cuts in federal environmental agencies (the USGS took a 22 percent hit) saying government—the EPA for instance—should get out of the way of the people. Industry leaders and lobbyists, e.g., from mining and timber interests, were appointed to head sensitive environmental programs in the Department of the Interior and the EPA while their scientists were pushed aside (Shulman 2006; Mooney and Kirshenbaum 2009; Mooney 2005; Weiss 2004).

Science in government, especially the work that underpinned regulations, was to suffer what some scientists considered a major blow when the Data Quality Act, was inserted into a huge appropriations bill in 2000 and quietly passed into law in 2001. The act, purportedly written by industry, directed the Office of Management and Budget to assure that all federal government information is "reliable." It has largely become a tool for industry to challenge government data—for example, data supporting a ban on the use of arsenic preservatives in playground equipment, information about the hazards of the metal nickel, or the data demonstrating the reality of global warming (Weiss 2004). In its wording, the act appears harmless; in practice it's been used to impede the development of regulations and other government actions that are based on scientific data.

In 2000, the frog investigation suffered another setback in staffing. That winter Sue Kersten left to begin married life in Wisconsin, and then, unexpectedly, our supervisor told Jeff Canfield he would have to leave in August, even though we had the funds to keep him for another year. What happened to that money? Losing Jeff stunned me: a wealth of experience with the frog sites and his intimate knowledge of the frog database went out the door with him. Local media cited agency cutbacks in the frog work (Meersman 2001a, 2001b). The future looked bleak.

To our dismay, my outstanding wetlands assistant, Cade Steffenson, who was busy identifying the invertebrates in our samples, was let go literally days before the promised conversion of his position to a well-deserved permanent job for which he'd competed against other candidates. Because of his temporary worker status, when he was seriously ill for a week, he not only had no health coverage, he received no salary.

Our work was drowning.

I mounted two black flags on dowels and hung them off the top of my filing cabinet to symbolize Mark's and my positions now on the agency's chopping block—and my drooping spirits. In the mornings as I approached the MPCA building I dreaded going inside. I hated myself for feeling that way, because the work was so important to me.

Unexpectedly, our section manager stopped by to see how Mark and I were faring. He gazed at the black flags and inquired about our morale, something he'd never done before.

"Well, last night I dreamed of trying to rescue a drowning woman and no one would come down a hill to help me with the CPR. It was awful," I said.

"I've been asked to think of where else you and Mark might fit in the agency," he said, speaking quietly. He said some of the other managers were privately concerned that they might lose their jobs if they confronted the commissioner over the staffing changes. One manager had already been demoted, he said. I sensed his deep disappointment in how things were going at the MPCA. In spite of his impatience with Mark and me because of the frequent, unpredictable events related to our work, I knew he valued the agency's ability to promote a clean environment. He was not happy. And when he retired a year later, he packed up his office and moved out over the Fourth of July weekend without a word. After thirty years of working at the MPCA and overseeing the early development of wastewater treatment plants across the state, he left suddenly. No chance for staff to say good-bye; no chance to give him a retirement party.

Then we had more bad news: the agency wanted to take an *additional* $50,000 out of our EPA wetlands grant, an action we pointed out was probably illegal.

Someone asked later how I persevered through the ups and downs at work. I responded by describing the time when my first husband was recovering from his stroke. He stood at the blackboard in a lecture hall with me sitting at the back as a "student" for his upcoming organic chemistry class. He said to me, "I know what I want to say, but I can't say it." He tried nonetheless to teach, rejecting the option of using his long-term disability insurance because doing so meant he could not teach in the future. "*That* was perseverance," I said. It affected me. From then on I fought to have a meaningful career, to help out.

In 2001, an article titled "MPCA Seen as Adrift, in Disarray," appeared in the Minneapolis *Star Tribune.* It slammed the agency, partly for its laxity on permits related to clean water and air quality, and partly for the current state of organizational dysfunction (Meersman 2001c).

"The MPCA used to be known as a protector of citizens from polluters. Now it seems to be protecting polluters from citizens," commented State Representative Jean Wagenius in the article. Representative Munger had died in 1999 at age eighty-eight, and Wagenius and State Senator John Marty were now the legislature's environmental conscience (Sweeney and Caple 1999).

By June of 2001 the MPCA had publicly offered $750,000 to consultants for yet another "structural redesign" of the agency, to develop "process re-engineering and change leadership training." We'd heard managers decry the outdated "command and control" approach to agency function. Did that mean leadership would succumb to the wishes of outsiders? Collaborating and forming partnerships and listening to the "stakeholders" sounded good; being "more flexible" and encouraging "voluntary compliance" to regulations made me wary. Might these approaches compromise environmental protection?

Organizations of conservatives were busy then writing articles against the "command and control" style of regulation by the government, against the idea that government tells people or businesses what they can and can't do, whether it involved regulating fisheries or discharges of pollution into air and water. Such groups touted the advantages of "self-regulation" and greater flexibility by government regulators. At the end of the decade, one think tank, the Cato Institute, was calling for Congress to repeal several important environmental laws, including all regulatory programs directed at wetlands preservation that derived from Section 404 of the Clean Water Act, eliminating "command and control" dictates of the Clean Water Act related to pollution discharges, and repealing other major laws that aimed at reducing pollution. Did such organizations influence the MPCA's commissioner and leadership? Encourage the agency to drop its Clean Water Act wetlands protection program?

One day I'd peeked into an empty managers' conference room at the MPCA and saw statements written on newsprint flip charts that said:

> What is our role and how do we fulfill it?
> Who do we serve?
> Are we guardians of the natural environment?
> Or are we mediators between the environment and various stakeholders?

Weren't these issues resolved decades earlier, when the MPCA had developed its mission? At the side of the room, a couple of large Tinker Toy structures stood on a table. I shuddered.

Had the managers completely lost their way? Did they have no control over the situation, have to play along with this stonewalling for fear of losing their jobs or being be demoted? Later that year, Doug Grow, a newspaper columnist, wrote: "The MPCA, once a national model of vision and regulation, has become expert only at reorganizing itself. The MPCA specializes in dazzling flow charts that show how the work it doesn't do would be done if it were doing it" (Grow 2001).

By June of 2001, citing budget constraints, the MPCA formally announced that it was ending the investigation into the deformed frogs (Lien 2001; Meersman 2001e). Immediately, I made plans to transfer all the preserved frogs that remained in our basement storage area to a museum, so they would not be accidentally discarded after I retired. Hearing in late May that the MPCA was about to bail out of the frog investigation, amphibian scientist Mike Lannoo had been quoted in the Minneapolis *Star Tribune,* as saying, "I do think deformed frogs are a problem. . . . It's scary. I think people should be concerned" (Meersman 2001d).

Shortly, Minnesota Public Radio ran a long piece by reporter Mary Losure on the demise of the frog investigation (Losure 2001). One of our federal research partners was quoted as saying, "It's discouraging. MPCA was a real catalyst for bringing everybody together." Another called us a focal point for the on-the-ground work. One scientist said that this was a bad time for the MPCA to pull out when more information from the field was needed. An anonymous person suspected the MPCA's managers were uncomfortable with the possibility that deformed frogs indicated a wider environmental problem in the state. A scientist from California stated there was strong evidence that chemicals caused the deformities, but, so far, the research had not yet narrowed down many of the possible agents, nor had it pinpointed whether mixtures of chemicals might be responsible.

The investigation into Minnesota's deformed frogs was over.

Ironically, the MPCA's commissioner, a woman who had previously worked for a gas company, said on Minnesota Public Radio:

"The Pollution Control Agency is not a research institution. We are a regulatory agency and our thrust is not research, and what you're asking is something that is outside the normal mission of the agency. We don't have staff on board with those skills."

By late August, after the June demise of the frog investigation, an MPCA staff person brought in over a hundred live young toads that he and his four-year-old had collected around a suburban pond. We took them surreptitiously to the basement biology lab, where I could examine them.

"Look at this," I said. "There are two size classes of juveniles here—the one-inch-long ones look pretty good, but these really tiny toadlets have deformities." Barely one-fourth of an inch long, they couldn't possibly survive, unless they grew rapidly before winter.

Using a microscope, I could see the whole gamut of malformations in miniature. Some lacked a leg but had a foot pressed against the body. Several had partial legs, or legs bent and shortened. A few had abnormalities in the front limbs, skin webbed across a back leg, or atrophied limbs. Of 163 tiny young toads, 20 percent were seriously malformed.

All I could do was to alert Dave Hoppe and a scientist in the local USGS office. Clearly, the frog problem hadn't gone away.

Only the MPCA had.

That same summer, the agency's commissioner proudly announced that the downsizing of the MPCA's staff had been completed sooner than expected. "We will use our one-time savings to improve the physical environment of our staff," she wrote in a memo. Older office furniture would be replaced, and a new sustainable building would soon be proposed.

"Our hope is to gain additional cost savings that can be returned to environmental programs and reside in a building which demonstrates our commitment to the environment," she said.

The MPCA had truly lost its way. We didn't need a new building! We needed a full staff empowered once again to carry out the agency's important environmental work, with a clear mission to protect healthy environments and living species in Minnesota, whether wetlands or streams, frogs or fish, air or land.

From where we sat, Mark and I could glance up from our computers and watch the demolition of an old processing plant across the street to prepare for the construction of a new police facility and prison for Ramsey County. The wrecking ball swung back and forth, and old brick walls crumbled to the ground. In the background we could see the tall buildings of St. Paul.

"How easy it is to bring a building down," I said. "It happens so fast."

Mark gazed out the window. "True." he said. "It's the rebuilding that takes a lot more work."

A few months before I retired in 2002, I spoke about the importance of wetlands and about the frog investigation at a college in St. Peter, Minnesota. Afterward, as I headed home along the Minnesota River, I decided to stop at the Ney Pond. I drove off the gravel road through a field of dried corn stalks to the pond that had hatched so many gruesome frogs, the pond whose frogs had engaged kids and teachers, media and scientists. Seeing November's tan fringe of shoreline grasses and the pond's low water and exposed muddy edge triggered a surge of emotions I hadn't expected.

I parked my car in the open weedy area where I had first photographed the horror of the deformed frogs that hot August day in 1995. This was the place where we had netted hundreds of frogs; where we sat down on coolers to measure the frogs and record their deformities in our yellow-covered field notebooks; where we spoke in low, somber tones and described the unbelievable malformations; the place where we talked with concerned students, parents, teachers, and reporters.

I walked up the slope, sat down on a new wooden bench someone had installed beside a large bur oak tree, and gazed across the silent pond. It looked so calm. What secrets did it hide that we would never uncover? We knew that pesticides flowed into the pond, but did they harm the frogs?

The day was warm for late fall; the frogs had long since migrated to the river, or died trying. A breeze blew up the slope, and dry grasses bent and scraped audibly against each other. A woodpecker muttered quietly as it hopped over the craggy bark of the bur oak's trunk. A strange sense of peace passed over me.

Then I saw a huge flock of white pelicans soaring over the Minnesota River basin, their coordinated wing beats catching the sun and flashing white. As they wheeled in a large spiral, they would almost disappear from sight until the next up-swing, when the sun's brilliance once again spotlighted their magnificent wings. I sat in silent awe and watched the synchronized aerial dance. Gradually the pelicans swirled out of sight up the river, leaving me with a quiet calm I'd not experienced all those times I had waded into that pond and chased frogs.

I reflected on the termination of the frog investigation: all that effort brought to a standstill; all those connections severed before the frog mystery was solved. I thought about the people who wanted us to figure out the cause, about our human fallibilities and the unfinished work. What will the future hold? How will frogs fare?

I leaned over and picked up a handful of bur oak acorns and thought about their hidden potential, about the new life these robust seeds represented, about the sturdy tree that resists prairie fires, its dark, rugged branches spreading over me. That magnificent oak, standing as a sentinel on the low hill by the pond, had arisen from a single seed, from just one acorn. I fingered the curly fringe around the caps and the smooth shiny sides of the brown nuts. A sense of hope warmed my hands.

New life will emerge. Teachers and parents will plant seeds in children and inspire them to go outside and *look* at nature. Fresh eyes will be opened to observe and report about the condition of the environment, about the health of wetlands and frogs. This will make all the difference, I believed. Another generation of researchers will guide efforts to restore frog habitats. Perhaps better regulations will be enacted to save wetland biodiversity. Scientists who work in government agencies will once again be empowered to speak the hard truth about difficult problems in the environment and propose ways to fix them. Agencies that regulate pollution and work to conserve life will become more proactive about environmental problems and support, not censor, their scientists.

I walked down the slope to my car, got in, and drove slowly past the new building constructed to house the Ney Nature Center. I turned onto the highway and headed home. It was November. By now the leopard frogs should be safely underwater in quiet backwater pools of the Minnesota River—if enough of them survived and made it down there from various shallow wetlands. Come spring, the river will warm and the frogs will crawl out onto land. On rainy April nights they'll hop across the two-lane highway and make the steep climb up the big bluff toward the thawed Ney Pond.

Males will growl their low, rumbling calls that are so peculiarly attractive to the females. Pairs of frogs will mate, and their fertile, jellied egg masses—containing two thousand or more eggs—will once again float near the pond's surface. Tiny tadpoles will hatch and grow rapidly, then will transform miraculously into juvenile frogs.

Then, I hoped and prayed, young frogs will emerge by midsummer from the pond—only this time with four healthy legs, two good eyes, and robust and healthy bodies.

One Sunday morning around that time, Verlyn was guest preacher for his former congregation, known for its social activism and inclusiveness. Verlyn and his predecessor both believed in preaching "with the Bible in one hand and the newspaper in the other." The church, housed in a historic brick structure, is tucked in beside tall medical buildings on the University of Minnesota campus. Verlyn's sermon topic, "Get Out of the Boat," was based on the biblical text where Jesus implores Peter to take risks and get out onto the water. With his usual, self-deprecating humor Verlyn had us laughing—and thinking. Aren't we like Peter, fearful we'll sink and drown if we leave the boat? take risks? In typical style he used some personal examples: his retirement and his moving on after painful life experiences. I knew by then that Verlyn felt as I did, that he and I had each come a long way toward finding peace with each other and with our lives. We'd both stepped cautiously out of our respective boats and risked a new life together.

As I sat listening, I reflected on my family, my life. At thirty-three, son Steve had just married and was settled into his job as a hydrogeochemist who worked on polluted landscapes. And Erik, now thirty-five, was about to graduate with his degree in architecture. The previous fall, in the evenings, I'd taught my newly created college course in pollution biology, and I was preparing to teach Biology of Women again, a course I'd taught and loved several times before coming to the MPCA. I'd finished the guide to identifying invertebrates for citizen volunteers. Our program for training people to monitor the biological health of local wetlands, the Wetlands Health Evaluation Project, or WHEP) was taking on a life of its own in two major metropolitan counties. I'd finally completed EPA's professional guide to assist states in surveying the invertebrates for monitoring wetlands water quality, a document written with several national wetlands scientists. In addition, that final year, I'd created and given workshops for schoolteachers on vernal pools at Audubon's woodsy environmental center north of the Twin Cities.

I was preparing to move on to a new life. But in my heart I knew that I'd never lose my deep concern about the future for frogs and wetlands.

EPILOGUE

HOW ARE THE FROGS? AN UPDATE

More than a decade has passed since the MPCA halted its research on the frogs in 2001 and after my retirement in 2002 to compose the next chapter of my life teaching, writing, and spending time with family. Since then, major environmental disasters, such as the catastrophic BP oil spill in the Gulf and growing concerns about climate change have rightly dominated the news. Media coverage about frogs, on the other hand, has been almost nonexistent. Yet scientific reports from the first decade of the twenty-first century are disquieting: biologists report significant numbers of malformed frogs from sites across the country; male frogs have female egg structures growing inside their testes; and frog populations continue to decline both nationally and globally. In many areas the loss and degradation of natural wetlands continues.

What does the future hold for frogs and the wetlands they need? Are we headed for another kind of silent spring, one with fewer frogs calling to attract mates?

The Current Status of Frogs

Globally, an astonishing one-third of amphibians (1,856 species) are classed as threatened with extinction. Among these, 427 species are listed as critically endangered, which means they are poised on the brink of extinction worldwide. And a whopping 43 percent, or 2,468 species of amphibians, are experiencing some degree of population decline (Stuart et al. 2004; see update in Alford 2010). In spite of these troublesome facts, research on declining amphibians receives less support relative to that for work on threatened birds and mammals (Lawler 2006). Are frogs and salamanders not charismatic enough to merit more attention? Of less economic consequence?

In the United States, twenty-three species of amphibians are classified as federally endangered or threatened, with eleven more waiting to be listed. Causes of these declines are multiple: habitat loss, diseases—particularly the deadly chytrid fungus that infects frog skin—and pollution present major threats. Invasive species and commercial exploitation also reduce amphibian populations (Alford 2010). The future for amphibians, both salamanders and frogs, appears bleak.

Even the northern leopard frog (*Rana pipiens,* now named *Lithobates pipiens*) may be in jeopardy. At one of our study sites I remember the carpet of newly metamorphosed leopard frogs that covered the shoreline so thickly we had to wade into the water to avoid squashing them with our boots. We surveyed more than twenty thousand young leopard frogs during the course of the Minnesota frog investigation. They were everywhere, or so it seemed.

Yet during the 1990s I heard stories from local people who no longer saw the mass migrations they'd experienced in their youth when frogs blanketed the roads near rivers. Harvests of leopard frogs in Minnesota to use as fish bait and for biological supply houses have continued, but records aren't kept. Bob McKinnell and Dave Hoppe surveyed frogs in Minnesota from 1967 into the 1990s. Before 1975, they collected almost twice the number of frogs with body lengths exceeding 60 mm compared with more recent surveys. By the 1990s, frogs they collected were smaller overall. And "lunkers," or frogs measuring 90 mm or more, were extremely rare (Hoppe and McKinnell 1997). "The frogs I'm seeing now are younger, they're not living as long as they used to," Bob told me.

More recent reports about the once common leopard frog are troubling: extirpated in two-thirds of its historic locations in the western United States, they are declining dramatically in the provinces of British Columbia and Alberta in Canada (Wilson et al. 2008). They've disappeared from 95 percent of their historic range in California, and most have disappeared from the states of Washington and Oregon (Lannoo 2005). In eleven states the leopard frog is classed as "critically imperiled" or "imperiled." That's about a third of the states where it's considered an endemic, or native, species. Populations are declining in Colorado, Indiana, Wisconsin . . . and the list goes on. I had no idea that the ordinary leopard frog was in such danger. How did this come about? I expected it would always be there, abundantly so. Small solace that in Minnesota *Rana pipiens* is considered "apparently secure."

Malformed Frogs

Researchers have shown that abnormalities in frogs have definitely increased in frequency and severity in the past twenty years (Johnson et al. 2010). They've analyzed records and museum collections of frogs from the mid-twentieth century and compared this information with contemporary surveys at the historic collection sites. In Minnesota, for instance, when Dave Hoppe resurveyed sites where zoologist Dave Merrell had collected frogs in the late 1950s and 1960s (Merrell 1969), he found more severe and widespread malformations at Merrell's sites in the 1990s. He also examined extensive, historic collections at the University of Minnesota's Bell Museum of Natural History and he found a similar pattern (Hoppe 2000).

In 2003 Pieter T. J. Johnson and others collected frogs from a pond where Pacific tree frogs had been surveyed in the late 1950s and early 1960s. They found significantly more malformations there in 1999–2002 than recorded earlier. More broadly, the authors resurveyed other sites, evaluated museum specimens, and analyzed data from the North American Reporting Center for Amphibian Malformations, or NARCAM. The qualitative evidence suggested that the number of frog populations affected has increased in recent years, a trend described as a "malformation epidemic" (Johnson et al. 2003).

In Arkansas the frequency of abnormalities in the northern cricket frog increased from 1957 to 2000 (McCallum and Trauth 2003). Using museum collections at the University of Arkansas, the authors documented 3 percent of frogs with abnormalities from 1957 through 1979 and 5.5 percent in the decade of the 1980s, and by 2000 8.6 percent of frogs were abnormal.

By the time MPCA staff conducted their final frog surveys in 2000, nearly 25,000 frogs had been examined and recorded. More than 2,000 of these had malformations—or 8.2 percent overall from 100 different ponds across Minnesota (data from Bob Murzyn, MPCA database; see also Vanden Langenberg et al. 2003). The 25,000 frogs surveyed by MPCA field crews includes those collected from wetlands targeted as reference sites, ponds that had few if any malformed frogs. By the time the investigation was ended in 2001, the MPCA had received several hundred reports of deformed frogs from citizens who lived in sixty-nine of the state's eighty-seven counties.

Approximately 60 percent of the malformations observed by MPCA staff consisted of limb deficiencies, principally missing or partial limbs and digits, plus a few with limbs bent at odd angles or rotated unnaturally. Frogs with extra or branched limb structures were uncommon, 9 percent overall. Less frequent malformations included webbed skin pulled tight across the legs, missing or abnormal eyes, atrophied limbs, misshapen jaws, and spiny posterior projections. One frog had an eye inside its throat. Most of the frogs collected in our studies (86 percent) were northern leopard frogs, but deformities were observed in a total of six species of frogs. Before the 1990s, Dave Hoppe told me, he'd rarely seen a malformed frog in more than 10,000 frogs he'd examined. The upsurge in deformed frogs in Minnesota during the 1990s marked a distinct change.

So, how are the frogs since the MPCA stopped surveying them in 2001?

In the summer of 2000 the US Fish and Wildlife Service (USFWS) began conducting surveys of frog populations for malformations in National Wildlife Refuges across the country. I am heartened that a federal agency took up the charge of checking on the status of young frogs; I am disheartened that they are finding more malformations. Some of the refuge studies, conducted over the past several years, have investigated potential causes of the deformities: parasites; contaminants in pond sediments, in water, and in frog tissues; and predators. Nationally, the Fish and Wildlife Service has analyzed frog populations in at least 152 refuges in 45 states across the country.

A ten-year review and summary of the findings by the USFWS in the refuges is forthcoming. The report will cover the work from 2000–2010 (USFWS 2011; Johnson et al. 2010). Preliminary data from these malformation surveys show that many sites had low rates of malformations in some years, high in others. Frequencies of malformations also varied by month and location. For example, surveys in Maryland showed low percentages of frogs malformed in 1999–2000 in three refuges (less than 3 percent). But in 2003 and 2004, 9–13 percent of southern leopard frogs, more than 15 percent of pickerel frogs, and 5.6 percent of green frogs were malformed in some of the USFWS collections in Maryland (Pinkney et al. 2007). In 2005 frog abnormalities above the background rate of 3 percent were recorded in refuges in Pennsylvania, New Hampshire, New Jersey, Vermont, and Maryland (Pinkney et al. 2006). In Alaska malformations averaged 6.2 percent in eighty-six breeding sites in five national refuges (Reeves et al. 2008). Rates of malformations ranged up to 20 percent in some of the Alaska survey sites.

The malformed frog phenomenon is widespread. In Bermuda, deformity rates as high as 15 to 25 percent were observed (Linzey et al. 2003). In the southeastern United States, USFWS staff have documented varying but sometimes high percentages of malformations on particular dates in refuges in Alabama (as high as 22 percent), Arkansas (up to 42 percent), Louisiana (24 percent), up to 16.8 percent in Mississippi, and up to 12 percent in sites in both North Carolina and Tennessee (Hemming et al. 2010). In New Hampshire, rates ranged up to maxima of 4.5, 9.8, and 15.4 percent in three different ponds in the Great Bay National Wildlife Refuge (Pinkney et al. 2006).

In 2011 Minnesota biologist Dave Hoppe found 10 percent deformed leopard frogs at a site in central Minnesota (Douglas County): 14 of 147 juvenile leopard frogs (10 percent) he surveyed had malformations. Most of them had missing or partial hind limbs, a few had vestigial (undeveloped) feet or toes at the knee joint. Yet in the late 1990s, Hoppe had collected hundreds of frogs at the same site and not one frog had a deformity (Hoppe 2011). Also in 2011, a family that lives a few miles from Henderson, Minnesota, found half of up to 40 young leopard frogs they collected near their pond had deformities: one or both feet missing, stumps of legs, one with an extra leg. They excavated their pond several years ago, and in recent years frogs are using it for reproduction. The problem of deformed frogs has not gone away.

Some conservatives have claimed that abnormal frogs have been around for a hundred years with "similar rates and kinds of deformities" (A. Avery 2004) and have suggested that deformed frogs are "simply another harsh fact of nature" (D. T. Avery 1999b). But scientists have found otherwise. One thing is clear: the malformations in frogs, while varying in time and place, have not abated. I am alarmed that young frogs continue to develop with defective limbs and other structures.

Parasites and Frogs

Frogs commonly harbor an array of adult and juvenile forms of parasitic worms, such as flukes and roundworms, in their tissues and organs. Different types of adult trematode flukes take up residence in the lungs, urinary bladders, or the upper and lower intestines. Immature stages of some species of flukes invade and embed themselves inside frogs' muscles and tissues, where they develop a protective, rounded case called a metacercarial cyst. These cysts are then consumed by frog-eating vertebrates such as herons, mammals, or snakes, where they develop into the mature, sexually reproducing flukes. The burden of parasites can be quite high. A single, normal green frog had over 1,200 cysts of the common trematode, *Fibricola,* in one limb's muscle tissues. The parasite, *Fibricola,* is not implicated as a cause of deformities (Gilliland and Muzzall 2002).

One biologist, the late Dan Sutherland, suggested that a diverse array of parasites in frogs might actually indicate a healthy ecosystem, since parasites, like other invertebrates, are vulnerable to pollution (Sutherland et al. 2002). Others have shown that increased pollution depresses the frog's immune system, making them more vulnerable to parasitic infections (Kiesecker 2002; Linzey et al. 2003; Rollins-Smith 2004).

Early on, parasites were proclaimed in the press as *the* solution to the mystery of the deformed frogs, even before any conclusive demonstration that the immature stages of the fluke *Ribeiroia* could cause malformations in developing tadpoles. The message, repeated especially in articles written by conservative writers (e.g., D. T. Avery 1999b; A. Avery 2004) was simple: the cause was natural! Why look at anything else?

I was skeptical then—parasites and frogs have coevolved and lived together for millennia. Why would one species of parasite be causing increasing rates of frog deformities all over the United States and in areas of Canada starting in the 1990s, when previously only sporadic incidences have been reported? Subsequently, however, lab studies did confirm that *Ribeiroia* could indeed trigger malformations (Johnson et al. 1999; Schotthoefer et al. 2003). Later, lab exposures to *Ribeiroia*'s immature stage resulted in malformations in the hind limbs, primarily skin webbing (cutaneous fusion) and bony triangles in American toads but not in tree frogs (Johnson and Hartson 2009). In Minnesota, only 4 percent of malformed frogs had bony triangles (anteverted limbs) and only 10 percent of those with deformities had webbed skin.

When an article was published in *Science* that demonstrated that *Ribeiroia* could trigger malformations in frogs (Johnson et al. 1999), conservative and libertarian groups unleashed a spate of communications. Some accused the media, environmentalists, and scientists, myself included, of "eco-suspicion" and "scaremongering" the public over a "frog fraud." We did this, in their view, to obtain funding and add staff so we could create more regulations to control what farmers and industry do. Government researchers, they claimed, wanted to blame industry for pesticides and increased ultraviolet radiation as causing the deformities in frogs. Global warming would become "the next culprit" for the "manmade frog catastrophe." I'm accused of having "Rachel Carson syndrome" because I publicly said we still needed to investigate chemicals, pesticides included, as possible causes of deformities, while looking at parasites and other factors.

Subsequently, the research of Schotthoefer, Koehler, Meteyer, and Cole (2003) has made it clear that the *Ribeiroia* parasite cannot explain the majority of deformities observed in frogs. In their experiments they observed duplicated and branched structures in limbs, feet, and digits of frogs exposed to *Ribeiroia.* But the parasite caused no limb deficiencies—no missing limbs, no partial limbs, and no missing digits.

The majority of malformations in frogs we collected in Minnesota consisted of reduced limb structures, namely missing or partial limbs (37.6 percent) and missing digits (20.7 percent), which accounted for 58 percent of the observed deformities. Other structural limb abnormalities included bent and rotated limbs (3.9 percent of total malformations). In contrast, duplications in limb structures (multiple or branched) represented less than 9 percent of the deformities in the Minnesota frog collections from 1995 to 2000.

In many other field surveys of frogs, limb deficiencies, not limb duplications and branches, comprised the majority of deformities recorded, e.g., in Ontario (Ouellet et al. 1997), in Vermont (Skelly et al. 2007; Levey et al. 2003), and in Alaska (Reeves and Trust 2008; Reeves et al. 2008).

In the federal National Wildlife Refuge studies, where overall 6.6 percent of frogs had deformities, more than 60 percent of these had missing or partial limb structures. Relatively few frogs had the type of abnormality attributable to the *Ribeiroia* parasite (Johnson et al. 2010 citing Guderyahn 2006; see USFWS refuge reports). In Alaska, for instance, more than one-half of the malformations recorded in several thousand frogs surveyed consisted of partial and missing limbs. Only one frog had an extra limb.

It appears that parasites, *Ribeiroia* in particular, cannot explain most of the malformations in frogs.

One extensively researched field study reported a significant relationship between the number of *Ribeiroia* infections in the Pacific coast tree frog and the frequency of malformations found in frogs in northwestern states (Johnson et al. 2002). But such relationships have not been observed in other studies. Researchers in northern Vermont, where malformed frogs were prevalent in the Lake Champlain basin, reported roughly 8 percent malformed frogs in seventeen of nineteen sites. Most of the frogs had missing and partial limbs, or digit deficiencies (Levey et al. 2003). Analysis of Vermont frogs at the USGS National Wildlife Health Center lab in Wisconsin showed no relationship between frogs with limb deformities and any kind of parasite.

And, more recently, Skelly (Skelly et al. 2007) presented further evidence that *Ribeiroia* could not have caused the malformed frogs in Vermont because no *Ribeiroia* cysts were found in a large sample of frogs from sites that had high frequencies of deformed frogs. Also, studies in southern Michigan (Gillilland and Muzzall 2002) concluded that parasites did not cause limb malformations there and *Ribeiroia* was not reported in frogs. Nor was *Ribeiroia* detected in any of the frogs surveyed in the Alaskan refuges (Reeves et al. 2008).

Dan Sutherland had researched *Ribeiroia* in malformed frogs over five years in Minnesota, Wisconsin, and northwestern Iowa from the late 1990s to

2001 (Sutherland et al. 2002). He and coworkers collected frogs from forty-four sites and examined them for the cysts of the *Ribeiroia* parasite. They found *Ribeiroia* at twenty-two sites, eighteen of them in Minnesota and in five species of frogs. Infection rates were sometimes high, averaging one hundred cysts per frog collected from three Minnesota "hotspots," that is sites known to have high rates of malformations. But *Ribeiroia* was not detected in frogs from other hotspots, especially those in the western part of the state. Several sites had high rates of malformed frogs and no *Ribeiroia.*

At the Ney Pond, Sutherland found *Ribeiroia* infection rates were low and sporadic in frogs, occurring in only two abnormal frogs with only one or two cysts each and also in some of the normal frogs. Yet one biologist speculated more recently that "trematode infections most likely caused the abnormalities" at the Ney Pond (a personal communication by P. Johnson cited in Collins and Crump 2009).

While Sutherland had found *Ribeiroia* in deformed mink frogs from a hotspot in Minnesota, Schotthoefer et al. (2009) detected no *Ribeiroia* infections in 144 mink frogs from thirty-six wetlands in other areas of the state (Schotthoefer et al. 2009), a collection with only 5 frogs malformed. The relationship between *Ribeiroia* and malformed frogs in Minnesota is unclear.

Biologists have suggested that increasing the fertility of pond water caused by increased runoff by phosphorus and nitrogen fertilizers can promote growth of snails that harbor the infectious stages of trematode parasites by stimulating more algae, eaten by snails. Experiments have supported this scenario (Johnson et al. 2007). Globally, the use of fertilizers has increased sharply in parallel with population increases. The levels of nitrate fertilizer have increased in rivers in direct relation to increases in population density (Keddy 2010) and in the United States, contributing to the expanding "Dead Zone" in the Gulf of Mexico. That I know of, however, there's little trend data that suggests the fertility of wetlands increased markedly starting in the 1990s across Minnesota.

What plausible explanations remain for the widespread surge in malformed frogs over the past fifteen to twenty years if parasites, at best, explain only a fraction of the deformities? Several researchers affirm the notion there may be multiple causes (or etiologies) of the deformities in frogs (Johnson and Bowerman 2010; Reeves et al. 2010; Lannoo 2008; Lannoo et al. 2003; Meteyer 2000, to name a few). And even today, several hypotheses remain in play, while new ones emerge.

Predators and Missing Limbs

Intriguing recent work that predatory invertebrates and fish can cause missing and partial limbs has raised a new, "natural" explanation for malformations in frogs. In one study, whose authors called the limb deficiencies "idiosyncratic," hungry dragonfly nymphs were placed in containers with young tadpoles and predation on the developing limbs was observed (Ballengée and Sessions 2009). Predatory invertebrates such as dragonflies, fish, and newts were collected in ponds that had malformed frogs. One of the authors, Stan Sessions, who has acknowledged that the most common deformities, limb deficiencies, are not caused by parasites, was interviewed on the BBC program *Earth News* (Walker 2009). There he claimed that selective predation by dragonfly nymphs "is now by far the leading explanation" [of missing and partial limbs]. Chemical pollutants are not necessary to understand either extra or missing limbs in frogs. I'd heard this years before, when he'd claimed that parasites were a "sufficient" explanation of frog deformities while saying the reason to analyze chemicals would be to allay human fears.

The suggestion that dragonfly predation could account for the majority of malformations in frogs across the United States seemed preposterous to some biologists. Nipping a limb bud couldn't explain the internal abnormalities in deformed frogs as described by pathologists from the National Wildlife Health Center Lab: loss of pelvis, deformed vertebrae, jaw and skull abnormalities, and partial notochord (see chapter 7). Predation doesn't explain the observation of immature (poorly calcified) bones in deformed frogs, or what Mike Lannoo referred to as "spongiform" bone. According to Lannoo, the spongiform bone that he observed in deformed frogs was "usually a terminal expansion of the cancellous [also called trabecular] bony matrix in missing limbs that cannot be associated with a recent predation event" (Lannoo 2011a; Lannoo 2008).

Some scientists point to the lack of field data to support the dragonfly hypothesis, explaining that no association had been shown between dragonfly densities and the numbers of malformed frogs in ponds (Skelly and Benard 2010; response by Sessions and Ballengée 2010). In long-term studies of the invertebrates and amphibians in thirty-seven wetlands located in the E. S. George Reserve in Michigan over twelve years, for instance, larval dragonflies were common, yet very few frogs, only 10 out of 36,151 surveyed, had limb malformations (Werner 2007).

How could predaceous dragonflies account for the surge in malformations seen in many states, including Minnesota? My work at the MPCA showed that dragonflies were sensitive to pollution; they declined sharply in the more degraded wetlands that we monitored (Gernes and Helgen 2002; Helgen 2002a). Higher species counts, we found, are an important indicator of healthy wetlands. More recent work at the MPCA confirms this (Genet and Olsen 2008). Dragonflies and developing amphibians have coexisted in Minnesota ponds for thousands of years. What could have changed during the 1990s? Had large numbers of wetlands suddenly become less polluted, fostering more dragonflies?

Another recent study links both dragonfly and stickleback fish predation to missing and partial limbs in developing frogs (Bowerman et al. 2010). These authors suggest that the limb abnormalities in western toads at one site in Oregon could be caused by predatory nipping of limbs by stickleback fish, and in Cascades frogs at another site by dragonfly nymphs. In ponds the researchers placed net-enclosed cages that contained tadpoles but no predators and observed that few limb abnormalities developed.

The story continues to unfold; the relationship between frog malformations and predators is complex at best. In National Wildlife Refuge studies in Alaska, malformations in frogs were positively correlated with the proximity of ponds to roads, used as a surrogate of human influences (Reeves et al. 2008). More recently, multiple stresses on frogs, both chemical and biological were evaluated (Reeves et al. 2010). In an experiment, wood frog tadpoles were isolated from predators in fiberglass screen exclosures in pond water, and limb malformations did not develop. Many sites had contaminants at toxic thresholds, particularly in the sediments, but the bag-style exclosure nets may have prevented tadpole contact with the bottom muds. Statistical analysis had shown that metals and certain organic contaminants in the sediments increased the chances of skeletal abnormalities by 16 percent and 40 percent, respectively.

But early-season abundance of dragonfly nymphs in the Alaska study was also positively related to the chances of frogs having skeletal malformations. How to explain this? Did excluding tadpoles from dragonflies or preventing tadpole contact with toxic sediments prevent the limb deformities? Was it dragonfly predation after all? The authors speculate that the retarded growth of tadpoles, a known effect from pollution, exposes them longer to predators and impairs their sensory ability to detect and escape from predators, the latter an effect of certain pollutants observed in fish. Tadpoles have a row of sensory cells along their sides, cells that help them orient and swim properly in the water. Pollution could impair a tadpole's ability to escape from a predator.

Chemicals and Malformations

For years I've felt that chemical pollution as a potential cause of widespread occurrences of frog deformities had been pushed aside in favor of natural explanations—like parasites and certain predators. We use many chemicals daily in various products. Some lodge in our body's tissues (CDC 2009), others get into the environment. Exempted from regulation under the Toxics Substances Control Act (TSCA) when Congress passed it in 1976 were 62,000 marketed chemicals; since the inception of TSCA 22,000 more chemicals have come on the scene, yet a mere 5 have been regulated under this law.

An abysmally small number of pesticides have water quality standards, which must be in place before they can be regulated in surface waters. In Minnesota, for instance, where hundreds of pesticides are currently used in the predominantly rural state, only thirteen pesticides have water quality regulations to protect aquatic life and twenty-four pesticides have standards in place for drinking water (MN Rules, chapters 7050 and 4717). Many of these are products (such as DDT and chlordane) that have long been banned by the EPA. Yet pesticides are found in a high percentage of rural wells and surface waters in Minnesota and throughout the Midwest. And, to repeat, little is known about the toxicity of the "inert" chemicals added to pesticide formulations.

Setting water quality standards for pesticides can be challenging. In recent years the MPCA had listed five streams as polluted, based on violations of its first standard for the herbicide acetochlor. But after agricultural chemical companies visited the agency and presented new findings, the agency loosened the acetochlor standard. As a consequence, the five streams were removed from the polluted waters list (Meersman 2008, 2007b).

More support is needed for government scientists to develop water quality and drinking water standards for hundreds of pesticides in active use. One hopes that the EPA will require better disclosure of the other chemicals added to formulated products and work to regulate other chemicals of concern. But can they succeed when powerful outside forces work to reduce government regulations? In addition, the public should have the right to know which pesticides or other chemicals—e.g., those used for gas exploration—are used near their homes and in the products they use. I was shocked that we could not find out what pesticides were used near the ponds with deformed frogs because the information was protected. Concealing information about chemicals that enter the environment by calling this "confidential business information" or "trade secrets" or "proprietary information" is not in the public interest. This does not protect the health of humans or other species.

The criticisms by conservative and libertarian writers that those of us investigating deformed frogs were scaremongering tapped into a much wider effort. Their organizations have worked hard with industries to cripple the government's ability to regulate chemicals, to reduce air and water pollution, to slow wetlands-destroying activities, or to screen products that might be harmful to human health (Weiss 2004; US House Report 2003; Mooney 2005; Mooney and Kirshenbaum 2009). The lack of water pollution standards for most pesticides suggests they're succeeding.

When I helped investigate the deformed frogs in Minnesota, I saw parallels between the malformed frogs and human birth defects. Images of thalidomide babies lacking both arms, of Vietnamese children purportedly malformed from Agent Orange, and photos of babies with cleft lips and palates hovered in my mind.

Birth defects cause heartbreak and lifelong difficulties for millions of families worldwide: 7.9 million infants are born every year, and only half or so survive past age five (Weinhold 2009). In the United States alone, 130,000 children are born each year with a major birth defect. Yet the causes of a stunning 60 to 75 percent of them are not understood even today (Brent 2004; Bale et al. 2003; Schardein 2000).

Some researchers suggest that only 5 to 10 percent of human birth defects can be attributed to environmental causes, a category that includes maternal exposures to particular chemicals, drugs, and infectious agents as well as risks from obesity, alcohol, and smoking. Genetic factors cause around 15 to 30 percent (Brent 2004; Bale et al. 2003). The high percentage of birth defects with unexplained causes leaves open the possibility that undetermined environmental factors could play an important role.

The emerging field of epigenetics raises the possibility that chemicals might cause developmental abnormalities that are heritable by altering how genes are expressed but not actually mutating the DNA. Of concern in this new research is the possibility that such changes, perhaps induced by endocrine-disrupting chemicals, might be stable over many generations (Crews and McLachlan 2006).

Several studies on human birth defects have linked the time of conception of babies in rural areas with the spring season when pesticides are most heavily applied to farm fields. In northwestern Minnesota, an area where wheat, sugar beets, and potato crops predominate, birth defects were greater in the children of pesticide applicators and more frequent in the population if babies were conceived in spring (Garry et al. 1996). In a follow-up study,

Garry and others confirmed the significant increase in birth defects in the children conceived in spring. Of the birth defects noted, 43 percent were musculoskeletal in nature (Garry et al. 2002).

In 2003 an EPA scientist found, in areas characterized by extensive acreage of wheat, higher rates of birth defects particularly in infants conceived in spring (Schreinemachers 2003). Another study showed elevated rates of birth defects in conceptions in spring that paralleled the higher concentrations of pesticides in the spring as measured by the USGS in surface waters nationwide (Winchester et al. 2009).

An intriguing recent study found greater numbers of birth defects in the limbs of babies whose mothers lived within five hundred meters of corn crops in Indiana (Ochoa-Acuna and Carbajo 2009). Only limb deformities (deficiencies, extra digits, clubfoot), not other types of birth defects, were significantly related to the mothers' proximity to cornfields. The main herbicide used on 90 percent of corn crops (but not on soybeans) in Indiana was atrazine. Research has shown male frogs developing ovarian tissue in their testes in response to low levels of atrazine (Hayes et al. 2011, 2010, 2003, 2002; Jooste et al. 2005). But no frog studies that I know of have demonstrated the development of limb deformities in frogs exposed to atrazine in the lab.

While such studies merely suggest associations, not causes, between agricultural practices and human birth defects, they clearly argue for further research and testing. One source reported that reproductive toxicity tests of pesticides with lab animals found that roughly 50 percent of tested herbicides could cause animal malformations, as could over 50 percent of fungicides and at least 33 percent of insecticides (Schardein 2000). In addition, 43 percent of metals tested in lab animals could cause birth defects. (For a review of developmental and reproductive toxic effects on humans by pesticides, see Iyer 2001 and Weselak et al. 2007). Clearly, more testing for developmental toxicity from exposures to such chemicals in humans and frogs is needed.

Chemicals and Frogs

Chemical pollution poses a great threat to frogs, along with habitat losses and diseases (Mann et al. 2009, citing IUCN information). Numerous studies have demonstrated lethal, sublethal, and developmental effects that a variety of chemicals have on amphibians, and I will only allude to a few of these here. *Sublethal* means that chemicals can harm long-term survival, growth, reproductive organs, and embryonic development while not killing the organisms outright. Chemicals can have direct or indirect effects. For instance, a variety of chemicals, pesticides included, depress the immune system, making frogs more vulnerable to infections or other agents and pollutants.

An important source for readers who want to explore the toxic effects of many types of chemicals on amphibians (and reptiles) in greater depth is the second edition of *Ecotoxicology of Amphibians and Reptiles* (Sparling et al. 2010). An overview chapter (Linder et al. 2010) summarizes a wide variety of laboratory and field tests on amphibians exposed to specific chemicals in recent years, studies that document toxic effects. Several other chapters go into different classes of chemicals in greater depth.

A variety of chemicals cause harm to amphibians. Work in the Fox River in Wisconsin has shown that greater amounts of PCBs in river sediments impaired the hatchability of frog embryos and reduced the number of species of frogs (Karasov et al. 2005). PCB oils, or polychlorinated biphenyls, were banned in 1977 but still persist in the environment. See Sparling (2010) for reviews of other organic contaminants that are toxic to amphibians, such as petroleum-generated organic chemicals and phenols. Several metals cause reproductive and developmental effects. Mercury and selenium and other metals are known to cause developmental malformations (see chapters 11 and 12 in Sparling et al. 2010a; Lemly 2008, 1999; USGS 2000). In one study, exposure to cadmium retarded growth in tadpoles and caused abnormal development in forelimbs and hindlimbs of frogs (Kashiwagi 2009).

In the Minnesota frog investigation, federal scientists identified several chemicals in pond water and extracts of pond sediments that caused malformations in frogs in lab tests (Burkhart et al. 1998; Fort et al. 1999a, 1999b). In addition, passing pond water through activated charcoal filters removed the ability of the water to cause deformities in early tadpoles. This work led the federal (NIEHS) scientists to suggest that whatever was causing malformations in frogs was in the pond water and its bottom mud. EPA scientists claimed that low concentrations of salts in one pond's water explained the deformities in the lab tests, but NIEHS contract scientists soon demonstrated that the salt concentrations did not cause the malformations. There was something else in the water (see chapter 9).

Later, USGS scientists collected fat-soluble chemicals with membranous collectors (SPMDs) from two ponds in Minnesota, one whose frogs showed high rates of malformations, the other a reference site. In lab studies, extracts from the membranous devices caused malformations in tadpoles (bony triangles in limbs and skin webbing), with the results differing depending on ultraviolet light exposures (Bridges et al. 2004). The researchers pointed out that the membranes excluded passage by parasites and infectious microbes.

Interestingly, these are the same types of malformations observed in lab exposure studies of the parasite *Ribeiroia* and American toads (Johnson and Harston 2009), which shows that both chemicals and *Ribeiroia* can cause similar effects. In addition, a chemical (retinoic acid) and the parasite can both cause branched limbs in frogs.

A number of studies have shown associations of elevated pesticide concentrations in agricultural areas and detrimental effects on amphibian populations (Lehman and Williams 2010; Mann et al. 2009), plus effects in laboratory tests as well. Field studies to assess the impacts of pesticides on frogs in the environment are fraught with challenges because frogs are usually exposed to a mixture of chemicals that can interact among themselves and with other factors. Ultraviolet light, for example, can either convert a chemical that could cause malformations into a less harmful form, or, alternatively, convert a harmless chemical into an agent capable of causing malformations (Bridges et al. 2004).

Malathion, an organophosphate insecticide, is the most widely and heavily used insecticide in the United States. In one study, malathion caused malformations in *Rana* tadpoles and caused a reduction in hatching success and viability of frog embryos (Budischak et al. 2008). Yet, when malathion kills the invertebrates that might prey on tadpoles, tadpoles may survive better.

Glyphosate (the active ingredient formulated in Roundup and related products) is the most heavily used herbicide in the United States. It has an enlarging global market attributable to the increasing use of crops seeds that are genetically engineered to resist its herbicidal effects. By itself, the active, plant-killing ingredient, glyphosate, is considered to be only moderately toxic to developing amphibians (Lehmann and Williams 2010). But other chemicals, whose toxicity is often not tested, are added to the commercial product before it's applied to crops.

For example, one formulated Roundup product contains a surfactant called POEA that is added to make the glyphosate penetrate through the cell walls of plants. This formulation has been shown to be acutely toxic to tadpoles at environmental concentrations in nine species of frogs (Relyea and Jones 2009; Lehman and Williams 2010). In addition, the tadpoles grew more slowly and were smaller at metamorphosis. Lengthening the tadpole phase exposes them longer to other chemicals, to infections and parasites in the water, and to predation.

A new study, prompted by reports of neural and cranial malformations in humans in Argentina, suggests that glyphosate by itself can impair development (Paganelli et al. 2010). Direct injection of glyphosate into the cells of frog and chick embryos (to bypass the need for a surfactant that promotes penetration into cells) caused developmental defects. Previously, glyphosate alone was considered relatively nontoxic. More research on this is clearly needed: what dose of the chemical was injected, and how can that be compared with concentrations of Roundup that might penetrate developing frog larvae in ponds?

Finally, the heavily used herbicide atrazine continues to generate controversy because of conflicting results on its effects on the reproductive organs of frogs (Bishop et al. 2010). As said, atrazine does not appear to cause limb malformations or other kinds of external deformities in frogs. But it is ubiquitous, water soluble, and easily absorbed by frogs through their pelvic patch, a vascularized, ventral area of skin through which frogs absorb water from wet soils and leaves, along with other pollutants that might be present. In surface waters and groundwater, atrazine is the most commonly detected pesticide, with a half-life of one to seven months (or longer—see Lenkowski 2008). More than half of streams surveyed in the United States by the USGS had atrazine in concentrations exceeding the EPA's "safe" threshold level of three parts per billion (ppb) in water (Bishop et al. 2010, citing Gillion et al. 2007).

Studies by scientists who exposed frogs to low levels of atrazine have shown several impacts: ovotestes, i.e., immature eggs in the testes of males (Hayes et al. 2011; Hayes et al. 2002; Hayes et al. 2003; Jooste et al. 2005), altered sex hormone levels, slowed development times, reduced body size, smaller muscles in the larynx, and depression in the immune system (Bishop et al. 2010). More recently, Hayes has observed complete sex reversal in genetically male frogs exposed to atrazine (Hayes et al. 2011; Hayes et al. 2010). Another study showed an increased frequency of intersex gonads in male toads collected from areas with greater exposure to agricultural landscapes (McCoy et al. 2008).

In addition, one study using high levels of atrazine found damage to the heart, kidney, and axial skeleton (Lenkowski et al. 2008). The presence of atrazine was a strong predictor of increased trematode parasite infections in declining leopard frog populations (Rohr et al. 2008a, 2008b). In another study, Skelly found more intersexed frogs in nonagricultural areas (Skelly et al. 2010), perhaps implicating other endocrine-disrupting chemicals.

Some studies had results conflicting with the research by Hayes and others on atrazine's effects on frogs (Bishop et al. 2010; Kloas et al. 2009). In 2004, Hayes criticized the researchers who claimed they couldn't replicate his lab's results with atrazine and frogs (Hayes 2004). In its review, the EPA declared that atrazine did not affect the sex organs in male frogs and further tests of atrazine were not needed (Steeger et al. 2007). Yet atrazine was banned fully in Europe in 2005, and its use stopped in British Columbia in Canada in 2007.

Even though atrazine is not implicated in the frog limb deformities, the controversies over this herbicide show the difficulties surrounding the regulation of pesticides in the United States. In 1998 the European Union adopted a regulation that restricted the use of a pesticide if it was measured in water at a concentration of one part pesticide per billion parts of water or greater (one ppb or more). In contrast, in the United States, significant harm must be demonstrated before a chemical's use can be prevented.

European regulatory policies take a more precautionary approach to overseeing pollutants. The precautionary principle is the idea that preventing or forestalling damage to health or the environment should be done even when strong proof of harm is lacking (Harremoës et al. 2002). The idea is to err on the side of safety.

In the United States, on the other hand, both politicians and industries work to require more stringent, if not impossible, burdens of scientific proof that any new and existing chemical causes serious damage before it can be regulated or, worst case for industry, removed from the market. Drug, energy, agricultural, and industrial chemical interests maintain these efforts on a large scale (Mooney 2005; Mooney and Kirshenbaum 2009; Shulman 2006; Gelbspan 2004, 1997). This apparent campaign to undermine the validity of established scientific research not only cripples the effectiveness of regulations, it endangers the health of both humans and the environment and profoundly diminishes the public's trust in science-based knowledge. One tactic is to keep delaying final assessments of chemicals until new research emerges (US GAO 2008); others include replacing qualified scientists on government advisory panels with industry representatives, censoring reporting by government scientists (UCS 2008), and giving minority research findings the same status as results agreed on by the vast majority of scientists (Gelbspan 2004, 1997).

Hormones and Frogs

Frogs and humans share similar hormones. For instance, the structure of thyroid hormone (TH) in frogs is identical to that of humans (Kashiwagi et al. 2009), and the manner in which TH regulates certain gene activity is quite similar at the molecular level. TH helps activate the expression of genes that promote animal development. The action of thyroid hormone can be disrupted (accelerated or stopped) by various chemical pollutants like PCBs, and chemicals used in plastics and other products—bisphenol A found in baby bottles and other plastics and phthalates used in making plastics softer—can disrupt thyroid hormone's action. More recently, most baby bottles are BPA free, yet the plastic continues to line many canned food products.

In frogs, thyroid hormone is essential for tail resorption, limb development, and other bodily changes when tadpoles metamorphose into young adults. In the human fetus, abnormally low levels of thyroid hormone in early development can reduce the growth of long bones, leading to dwarfism and causing retardation if not corrected.

During the Minnesota frog investigation, federal scientists (NIEHS) isolated thyroid-disrupting chemicals from pond water known to cause deformed frogs. Adding thyroid hormone to the pond water in their tests with tadpoles prevented the development of malformations.

The suggestion that thyroid hormone-disrupting chemicals might play a role in causing frog malformations needs further exploration. To date, much of the work on endocrine-disrupting chemicals has been related to chemicals that are estrogenic.

Globally, the use of nitrogen-based fertilizers has soared, with eighty-three million tons applied to agricultural lands. Runoff from fields and yards can transport excess nitrates into wetlands, shallow wells, and drinking water supplies (Guillette and Edwards 2005). Consumption of nitrates in water and food has been linked to increased thyroid cancer in women (Ward et al. 2010), altered uptake of iodine, and impaired development in frogs. Nitrates are emerging as a likely endocrine-disrupting chemical, especially of thyroid hormone functions, but they may disrupt male sex hormones as well. More research is needed, especially if the drinking water standard for nitrate should be tightened.

Recently the EPA has begun to develop methods for screening chemicals for thyroid-disrupting activity by exposing larval tadpoles and tracking the development of their thyroid glands and hormone levels (Tietge et al. 2010; Hornung et al. 2010; Zhang et al. 2006; Tietge et al. 2005). More work similar to this will be needed, plus more screening and regulation of existing and new chemicals that come on the market. Tests should be routinely made for a variety of endocrine-disrupting effects by exogenous chemicals on sex hormones, thyroid hormone, and adrenal and pituitary gland hormones, all essential for normal body functions. The next step: monitoring surface waters for thyroid-disrupting chemicals, similar to current monitoring for estrogen disrupters (Ferry et al. 2010).

Frogs and Humans

Scientists debate whether the losses of frogs signal trouble for humans, and that debate will continue. In some ways the futures of frogs and humans are wrapped together because we share common needs and respond in similar ways to harmful agents even though our habitats vastly differ. Both frogs and humans require contaminant-free foods, a space to live safely, and a clean environment in order to survive and develop normally. Frogs and humans share some common, basic biological processes that have been evolutionarily conserved. We share genetic and skeletal structures. We have similarities in certain hormones, in some developmental processes, and even in our immune systems.

When I read the work on the immune system in frogs (Rollins-Smith et al. 2004; Rollins-Smith 1998; see also Rollins-Smith et al. 2011a; Rollins-Smith et al. 2011b), I was reminded how complex and evolved these humble creatures really are. Frogs have an immune system with several components that are found in mammals: T and B lymphocytes (white cells), killer cells, and the ability to make antibodies. Even tadpoles can produce antibodies. The immune system of frogs differs from that of mammals in the unusual die-off of 40 percent of their white cells right at the time of metamorphosis from tadpole to juvenile frog. This heightens their vulnerability both to infections and to hazardous chemicals during the critical transition to the adult form. Once the tadpole has transformed into a young adult frog, its immune system redevelops. Amphibians secrete antimicrobial coatings from specialized glands in their skin, and this provides some defense against disease organisms (Rollins-Smith et al. 2011a; Rollins-Smith et al. 2011b).

The body of research that seeks to connect concerns about environmental influences on the immune systems of frogs with their increased susceptibility to infections will expand in the future (Gendron et al. 2003; Christin et al. 2003), partly because the deadly chytrid fungus is devastating many frogs.

Habitat loss is considered a (if not *the*) major cause of global amphibian declines. Many species of frogs in the United States require wetlands for reproduction, and these habitats remain under assault. One could argue that humans also need wetlands, if not for the biological diversity they support, then at least to protect them against flooding and to contain landscape runoff.

In 2001 the US Supreme Court dealt the vast number of isolated wetlands a serious blow by essentially removing them from the Clean Water Act Section 404 protection and regulatory oversight. At issue was the old, "navigable waters" language that derived from the historic origin of this part of the act, its origin in the 1899 Rivers and Harbors Act, whose goal was to prevent obstacles to navigation.

In June of 2006 isolated wetlands took another hit in the Supreme Court. In order to have protection under Section 404 of the Clean Water Act, the wetland had to have a "significant nexus" (or connection) to a navigable water. This is a potential disaster for many wetlands that frogs need for successful reproduction and for other wildlife as well. In 2007 Rep. John Dingle (MI), one of the original sponsors of the 1972 Clean Water Act, cosponsored a bill to reaffirm that the CWA applies to all waters of the United States. Representative Dingle referred to the recent US Supreme Court decisions that supported removal of isolated wetlands from CWA protections as a "bungle" (Pittman and Waite 2009, p. 299). Because of these rulings, isolated wetlands may no longer be regulated by the US Army Corps of Engineers or by local pollution control agencies unless states have developed laws to override this. States may be leery to investigate pollution in isolated wetlands (a type that predominates in the Midwest) because of these rulings.

The legal process that allows dubious replacements of natural wetlands that are destroyed for development has resulted in enlarged and deepened ponds that receive the water drained from many shallow wetlands. These flooded ponds are not suitable habitat for frogs, many of which need fishless, shallow wetlands to safely reproduce, sites where their tadpoles aren't consumed by predatory fish. In addition, wetlands are used by resource agencies to rear game fish and by bait dealers to grow sucker minnows. Excavation of shallow wetlands to deepen them for residential fish ponds, a common practice, goes unregulated. Without better safeguards, both wetlands and frogs remain in peril.

In the most recent report on status and trends of wetlands from 2004 to 2009 by the US Fish and Wildlife Service (Dahl 2011), the acreage of freshwater forested wetlands declined steeply, largely because of silviculture practices, which have little oversight by Clean Water Act regulations. Although freshwater emergent marshes showed a slight increase in area during this time period, the gains came mostly from created and re-established wetlands, whose ecological value continues to be debated. In the prairie pothole swath across the midwestern United States, freshwater wetlands showed a net loss. In this region, wetlands previously set aside in conservation programs are being re-cropped because of recent higher prices for corn and soybeans. This along with improved drainage, has resulted in a greater loss of wetlands. And al-though the newly delineated category called "freshwater ponds" has shown a net increase from 2004, natural ponds, comprising one-third of this class, declined slightly. In the same period, the acreage of urban and agricultural freshwater ponds, of questionable ecological value, increased about 15 percent.

This new report on the status of wetlands, which does not assess wetland water quality and condition, shows that the loss of acreage among the types of wetlands favored by frogs for reproduction—forested, emergent, and natural ponds—continues.

Under Section 303(d) of the Clean Water Act, the EPA requires states to identify surface waters that are polluted (or "impaired"). To meet this requirement, states monitor the condition of waters, traditionally rivers, more recently lakes, and in, very few cases, wetlands. Impaired waters are those that violate state water quality standards, e.g., for the level of dissolved oxygen needed in streams to support fish, for specific concentrations of toxic chemicals, or for levels of coliform bacteria that exceed the criteria for human health safety as defined in state standards. Once a water is identified as polluted and placed on the impaired waters list, the EPA expects state agencies to proceed with a plan to reduce pollution and ultimately restore water quality, at least in high priority waters.

Except in rare cases, wetlands identified as polluted waters have not been listed on the EPA's impaired waters list. As of 2008 thirty-seven states lacked water quality standards for wetlands; some states have yet to define wetlands as legal waters of the state, even though wetlands are part of the nation's waters in the Clean Water Act. You can't begin to protect a wetland from water pollution if it's not officially a state body of water. According to the EPA, only eight states have state government programs to monitor wetlands (Minnesota is one). The EPA expects soon to release its first National Wetland Condition Assessment report as a beginning attempt at tracking trends in the condition or quality of wetlands. Biological indicators to be used are wetland plant diversity and algae (see discussion in Dahl 2011).

In 1993 the MPCA moved to include wetlands in its water quality regulations, giving them aquatic life use status (Class 2 waters), similar to that covering rivers and streams. In a recent MPCA study of a small river basin in an agricultural region of Minnesota, a stunning 91 percent of the large depressional wetlands surveyed were rated as impaired (i.e., polluted) using well-established biological monitoring methods (Genet and Olsen 2008). But only a handful of a large number of the wetlands documented as polluted by the MPCA have officially made it on the "impaired waters" list required under the Clean Water Act. One explanation for this was, purportedly, that Minnesota has too many wetlands to start listing those deemed to be polluted and it's hard to create a clean-up plan for wetlands. Is the agency steering clear of listing isolated wetlands as polluted because of the Supreme Court rulings? Can polluted wetlands be listed only if they have a "significant nexus" to state-listed polluted streams and lakes? But not if they stand alone?

Wetlands remain the poor cousins to other state waters, even today.

There are a few positive signs. Some communities excavate holding ponds for storm water runoff so it doesn't go directly into natural wetlands. Others are requiring or planting wider buffer zones around wetlands that help reduce runoff. Many programs encourage wetlands restoration (see below), and state and federal agencies have set aside reserves with wetlands as wildlife management areas and waterfowl production areas.

What Future for Frogs?

Frogs emerged onto the land from their watery origins long before the most primitive forms of humans had begun to evolve, long before humans domesticated plants and animals and learned to irrigate crops and preserve food. Frogs were evolving long before humans began to drain wetlands, fragment and pollute frog habitats, and create 82,000 chemicals that could enter the environments of Earth. Many kinds of frogs need to return to water to reproduce, a life history trait that greatly adds to their vulnerability.

Amphibians have been evolving for 365 million years. Fossils of frogs appear in deposits around 170–186 million years ago and fossils of tadpoles around 140 million (Carroll 2009). When did frogs evolve the ability to call to their mates? Bony structures that support the larynx are found in ancient fossil frogs, leading Robert Carroll (2009) to suggest that frogs may have been calling in spring for many millions of years. Primitive humans, on the other hand, began walking on earth a mere 190,000 years ago.

Frogs apparently survived the major glaciations in North America better than other species such as mammals, perhaps because they are cold blooded and cold tolerant. But this does not mean that frogs will be equipped to survive the human-caused warming of the climate that will accelerate during the next century and destroy shallow wetlands. The changes that led to the glaciations in northern North America were by and large more gradual, so frogs had time to adapt and to migrate. In a rapidly warming climate the more mobile species—like birds and mammals—will be better equipped to escape from dried or frozen habitats than small frogs, which can travel only short distances each year.

In the Midwest many frogs gravitate to isolated, depressional wetlands, some of which are called "prairie potholes." These wetlands support a high percentage of the nation's duck populations as well as frogs and other bird life. Because their watersheds are often small, prairie pothole wetlands are considered under the greatest threat from global warming and reduced local rainfalls (Mitsch 2007) and their continued decline has been recently documented by the US Fish and Wildlife Service (Dahl 2011).

Through the course of my work at the MPCA and during this review of some of the more recent research on deformed frogs, my feelings of awe for the elegant and intricate lives of these creatures have deepened. I've learned how complex the interactions are within aquatic communities and among the diverse types of agents that can harm living organisms. Solving the causes of malformations and population declines in frogs is no easy task, let alone preventing them in the future. In spite of great gains made in our knowledge about the needs and sources of harm to amphibians, much more work needs to be done. Similarly, greater efforts should be focused on restoring and saving the nation's remaining natural wetlands.

What hope is there for frogs? Carroll describes amphibians as "tough survivors." He hints they may outlive us. Maybe so. But as I see it, we share the same Earth with them, we share environments that continue to be degraded. It is up to us to change course and promote a more healthy and sustainable world. The frogs can't do this. They need our help.

I am awed by many dedicated people who are out there "on the front lines." Mike Lannoo, as one among many, works passionately to protect amphibians (Lannoo 2011b; Lannoo in press) by writing, teaching at a field station, and restoring habitats of threatened frogs. Ed Little and other scientists at the USGS Columbia Environmental Research Center continue their research on the effects of ultraviolet light and pollutants on wildlife. Carol Meteyer and David Green at the National Wildlife Health Center Lab, Louise Rollins-Smith, and others work long hours seeking ways to understand diseases and parasites that may be contributing to amphibian declines.

Heroic biologists in zoos, universities, and government agencies are nurturing threatened species of frogs and attempting to release their fertilized eggs back to protected, natural habitats (Landis 2011). In addition, a number of scientists have dedicated their lives to understanding amphibian development at the very basic level. These and many other researchers are the amphibians' unsung heroes.

I'm also inspired by people like retired elementary school teacher Art Straub, who lives in Le Sueur, Minnesota. He and his wife, Barb, devote their energy and passion to teaching kids and adults to cherish the values of natural habitats and the life in the Minnesota River basin: the birds, frogs, and plants. Art opens people's eyes—and their thinking—while he keeps constant watch on the area's wildlife, like nesting eagles and migrating frogs. No surprise that the Straubs have been restoring tall grass prairie and forested areas on a couple hundred acres of land they own near the river. They are helping the land heal itself, Art says. I have singled him out, but people with Art's level of devotion to the environment—teachers, nature center staff, and ordinary citizens—abound. They're just quiet (as he is) about their accomplishments.

Nationally, dedicated volunteers in twenty-one states are keeping tabs on local species of frogs by going out after sunset and stopping at several assigned locations to listen to the calls of frogs. They use protocols originally developed in the Midwest to collect and report their data to the North American Amphibian Monitoring Program in partnership with the USGS (Weir et al. 2009). In addition, according to the EPA, in roughly fifteen states, including Minnesota, citizen volunteers are monitoring local wetlands.

Untold numbers of people, like the Straubs, are faithfully restoring wetlands and upland habitats. Many are encouraged and guided by a variety of publicly and privately funded programs, such as the Wetlands Reserve Program (USDA Natural Resources Conservation Service) and the US Fish and Wildlife Service's Partners for Fish and Wildlife Program, or by non profit organizations, such as Ducks Unlimited, Pheasants Forever, the Nature Conservancy, and the Izaak Walton League.

I applaud the host of unpaid volunteers who work to promote a healthy environment and better conservation of habitats through various nonprofit organizations (like Sierra, NRDC, Audubon), the writers and artists (as in *Orion* and *OnEarth* magazines) who speak the truth, and—dare I say it—the politicians, who, like Minnesota's late Rep. Willard Munger, consistently work for a safer, cleaner environment. The way forward lies in the hands of the many dedicated scientists, teachers, kids, government workers, and ordinary citizens who want to protect amphibians and help save our remaining, fragile wetlands before it's too late. They promote solutions and inspire us to keep trying, even when positive change seems out of reach. They believe otherwise.

To all this cloud of people who pour their lives and their passions into understanding and saving earth's vulnerable creatures and fragile habitats even when things look dire, I say: Godspeed.

REFERENCES

Alberts, Bruce. 2005. *AAAS Policy Brief: Access to Data.* Feb 10, 2005, Update and Testimony by the President of the National Academy of Sciences, Dr. Bruce Alberts, July 15, 1999. Washington, DC: American Association for the Advancement of Science.

Alford, Ross A. 2010. "Declines and the Global Status of Amphibians." In Sparling et al., eds., *Ecotoxicology of Amphibians and Reptiles,* 13–45.

Andrews, Richard N. L. 2006. *Managing the Environment, Managing Ourselves.* 2nd ed. New Haven: Yale University Press.

Ankley, Gerald T., J. E. Tietge, G. W. Holcombe, D. L. DeFoe, S. A. Diamond, K. M. Jensen, and S. J. Degitz. 2000. "Effects of Laboratory Ultraviolet Radiation and Natural Sunlight on Survival and Development of *Rana pipiens.*" *Canadian Journal of Zoology* 78:1092–1100.

Avery, Alex. 2004. *Rachel Carson Syndrome: Jumping to Pesticide Conclusions in the Global Frog Crisis.* Washington, DC: Hudson Institute, Center for Global Food Issues.

Avery, Dennis T. 1999a. Editorial, "Leaping to Conclusions about Deformed Frogs" *Global Food Quarterly* 27 (Spring): 2–3.

———. 1999b. "Deformed Frogs Another False Alarm: Once Again, Eco-suspicion Has Run Far Ahead of Science and Embarrassed Itself." May 11. Washington, DC: Hudson Institute. http://rs.hudson.org/index.cfm?fuseaction=publication_details&id=312.

Bale, Judith R., B. J. Stoll, and A. O. Lucas, eds. 2003. *Reducing Birth Defects: Meeting the Challenge in the Developing World.* Washington, DC: National Academies Press.

Ballengée, Brandon, and S. K. Sessions. 2009. "Explanation for Missing Limbs in Deformed Amphibians." *Journal of Experimental Zoology, Part B: Molecular and Developmental Evolution* 312B:770–79.

Bech-Danielsen, Anne. 1996. "Houston Vi Har et Problem." *Politiken* (Copenhagen, Denmark). Oct. 28.

Bishop, Christine A., T. V. McDaniel, and S. R. de Solla. 2010. "Atrazine in the Environment and Its Implication for Amphibians and Reptiles." In Sparling et al., eds., *Ecotoxicology of Amphibians and Reptiles,* 225–59.

Borkin, Leo J., and M. M. Pikulik. 1986. "The Occurrence of Polymely and Polydactyly in Natural Populations of Anurans of the USSR." *Amphibia-Reptilia* 7 (3): 205–16.

Bowerman, J., P. T. J. Johnson, and T. Bowerman. 2010. "Sublethal Predators and Their Injured Prey: Linking Aquatic Predators and Severe Limb Abnormalities in Amphibians." *Ecology* 9 (1): 242–51.

Bowerman, W. W., IV, T. J. Kubiak, J. B. Holt, D. L Evans, R. G. Eckstein, C. R. Sindelar, D. A. Best, and K. D. Kozie. 1994. "Observed Abnormalities in Mandibles of Nesting Bald Eagles *Haliaeetus leucocephalus*." *Bulletin of Environmental Contamination and Toxicology* 53:450–57.

Breckenridge, Walter J. 1944. *Reptiles and Amphibians of Minnesota.* Minneapolis: University of Minnesota Press.

Brent, Robert L. 2004. "Environmental Causes of Human Congenital Malformations: The Pediatrician's Role in Dealing with These Complex Clinical Problems Caused by a Multiplicity of Environmental and Genetic Factors." *Pediatrics* 113 (4): 957–68.

Bridges, Christine, E. Little, D. Gardiner, J. Petty, and J. Huckins. 2004. "Assessing the Toxicity and Teratogenicity of Pond Water in North-Central Minnesota to Amphibians." *Environmental Science and Pollution Research* 11 (4): 233–39.

Budischak, Sarah, L. K. Belden, and W. A. Hopkins. 2008. "Effects of Malathion on Embryonic Development and Latent Susceptibility to Trematode Parasites in Ranid Tadpoles." *Environmental Toxicology and Chemistry* 27 (12): 2496–2500.

Burkhart, James G., J. C. Helgen, D. J. Fort, D. Gallagher, D. Bowers, T. L. Propst, M. C. Gernes, J. Magner, M. D. Shelby, and G. Lucier. 1998. "Induction of Mortality and Malformation in *Xenopus laevis* Embryos by Water Sources Associated with Frog Deformities." *Environmental Health Perspectives* 106:841–48.

Calfee, Robin D., E. E. Little, L. Cleveland, and M. G. Barron. 2000. "Photoenhanced Toxicity of a Weathered Oil on *Ceriodaphnia dubia* Reproduction." *Environmental Science and Pollution Research* 6:207–12.

Capel, Paul D., M. Lin, and P. J. Wotzka. 1998. *Wet Atmospheric Deposition of Pesticides in Minnesota, 1989–94.* US Geological Survey Water-Resources Investigations Report 97-4026. In cooperation with the MN Department of Agriculture. USGS, Mounds View, MN.

Carroll, Robert. 2009. *The Rise of Amphibians: 365 Million Years of Evolution.* Baltimore: Johns Hopkins University Press.

Carson, Rachel. 1962. *Silent Spring.* Boston: Houghton Mifflin / Cambridge: Riverside Press.

Centers for Disease Control. 2009. *Fourth National Report on Human Exposures to Environmental Chemicals.* [and updated tables Feb. 2011]. Atlanta: Centers for Disease Control.

Christiansen, James L., and Heather Feltman. 2000. "A Relationship between Trematode Metacercariae and Bullfrog Limb Abnormalities." In Kaiser et al., eds., *Investigating Amphibian Declines,* 79–85.

Christin, Marie-Soleil, A. D. Gendron, P. Brousseau, L. Menard, D. J. Marcogliese, D. Cyr, S. Ruby, and M. Fournier. 2003. "Effects of Agricultural Pesticides on the Immune System of *Rana pipiens* and on Its Resistance to Parasitic Infections." *Environmental Toxicology and Chemistry* 22 (5): 1127–33.

Cobb, Kim. 1997. "Internet Feeds Concern about Frog Deformities." *Houston Chronicle,* Oct. 3.

Colborn, Theo, D. Dumanoski, and J. P. Myers. 1996. *Our Stolen Future.* New York: Dutton.

Collins, James P., and Martha L. Crump. 2009. *Extinction in Our Times: Global Amphibian Decline.* Oxford: Oxford University Press.

Corbet, Philip S. 1999. *Dragonflies: Behavior and Ecology of Odonata.* Ithaca, NY: Comstock.

Cowardin, Lewis M., V. Carter, F. C. Golet, and E. T. LaRoe. 1979. *Classification of Wetlands and Deepwater Habitats of the United States.* Washington, DC: US Fish and Wildlife Service.

Crews, David, and John A. McLachlan. 2006. "Epigenetics, Evolution, Endocrine Disruption, Health, and Disease. *Endocrinology* 147 (6): 4–10.

Crump, Marty. 2000. *In Search of the Golden Frog.* Chicago: University of Chicago Press.

Cushman, John H. 1997. "EPA Withdraws Plans to States Flexibility on Rules." *New York Times,* Mar. 2.

Dahl, Thomas E. 2011. *Status and Trends of Wetlands in the Conterminous United States 2004 to 2009.* Washington, DC: US Fish and Wildlife Service.

———. 2006. *Status and Trends of Wetlands in the Conterminous United States 1998 to 2004.* Washington, DC: US Fish and Wildlife Service.

Dawson, Doug A., B. D. Scott, M. J. Ellenberger, G. Poch, and A. C. Rinaldi. 2004. "Evaluation of Dose Response Curve Analysis in Delineating Shared or Different Molecular Sites of Action for Osteolathyrogens." *Environmental and Toxicological Pharmacology* 16:13–23.

Dawson, Doug A., M. A. Cotter, D. L. Policz, D. A. Stoffer, J. P. Nichols, and G. Poch. 2000. "Comparative Evaluation of the Combined Osteolathyritic Effects of Two Nitrile Combinations on *Xenopus* Embryos." *Toxicology* 147:193–207.

Dawson, Doug A., T. W. Schulz, L. L. Baker, and A. Mannar. 1990. "Structure-activity Relationships for Osteolathyrism: III. Substituted Thiosemicarbazides." *Journal of Applied Toxicology* 10:59–64.

Dich, Jan, S. H. Zahm, A. Hanberg, and H-O. Adami. 1997. "Pesticides and Cancer." *Cancer Causes and Control* 8 (3): 420–43.

Dodson, Stanley, and R. A. Lillie. 2001. "Zooplankton Communities of Restored Depressional Wetlands in Wisconsin, USA." *Wetlands* 21 (2): 292–300.

Ehrlich, Paul R., and Anne H. Ehrlich. 1996. *Betrayal of Science and Reason: How Anti-Environmental Rhetoric Threatens Our Future.* Washington, DC: Island Press.

Ferrey, Mark, H. Schoenfuss, R. Kiesling, L. Barber, J. Writer, and A. Preimesberger. 2010. *Statewide Endocrine Disrupting Compound Monitoring Study 2007–2008.* St. Paul, MN: Minnesota Pollution Control Agency [with St. Cloud State University and USGS].

Flax, Nina L., and Leo J. Borkin. 1997. "High Incidence of Abnormalities in Anurans in Contaminated Industrial Areas (Eastern Ukraine)." In *Herpetologia Bonnensis,* edited by W. Bohme, W. Bischoff, and T. Ziegler, 119–223. Bonn, Germany: Societas Europae Herpetologicas.

Fort, Doug J., T. L. Propst, E. L. Stover, J. C. Helgen, R. B. Levey, K. Gallagher, and J. G. Burkhart. 1999a. "Effects of Pond Water, Sediment, and Sediment Extracts from Minnesota and Vermont, USA, on Early Development and Metamorphosis of *Xenopus.*" *Environmental Toxicology and Chemistry* 18 (10): 2305–15.

Fort, Doug J., R. L. Rogers, H. F. Copley, L. A. Bruning, E. L. Stover, J. C. Helgen, and J. G. Burkhart. 1999b. "Progress Toward Identifying Causes of Maldevelopment Induced in *Xenopus* by Pond Water and Sediment Extracts from Minnesota, USA." *Environmental Toxicology and Chemistry* 18 (10): 2316–24.

Fox, Glen A. 1991. "Practical Causal Inference for Ecoepidemiologists." *Journal of Toxicology and Environmental Health* 33:359–73.

Garry, Vincent F., M. E. Harkins, L. L. Erickson, L. K. Long-Simpson, S. E. Holland, and B. L. Burroughs. 2002. "Birth Defects, Season of Conception, and Sex of Children Born to Pesticide Applicators Living in the Red River Valley of Minnesota, U.S.A." *Environmental Health Perspectives* 110 (suppl. 3): 441–49.

Garry, Vincent F., D. Schreinemachers, M. E. Harkins, and J. Griffith. 1996. "Pesticide Appliers, Biocides, and Birth Defects in Rural Minnesota." *Environmental Health Perspectives* 104 (4): 394–99.

Gatto, Nicole M., M. Cockburn, J. Bronstein, A. Manthripragada, and B. Ritz. 2009. "Well-water Consumption and Parkinson's Disease in Rural California." *Environmental Health Perspectives* 117 (12): 1912–18.

Gelbspan, Ross. 2004. *The Boiling Point: How Politicians, Big Oil and Coal, Journalists, and Activists Are Fueling the Climate Crisis—and What We Can Do to Avert Disaster.* New York: Basic Books.

———. 1997. *The Heat Is On: The Climate Crisis, the Coverup, the Prescription.* New York: Perseus Books.

Gendron, Andrée D., D. J. Marcogliese, S. Barbeau, M.-S. Christin, P. Brousseau, S. Ruby, D. Cyr, and M. Fournier. 2003. "Exposure of Leopard Frogs to a Pesticide Mixture Affects Life History Characteristics of the Lungworm *Rhabdias ranae.*" *Oecologia* 135:469–76.

Genet, John A., and A. R. Olsen. 2008. "Assessing Depressional Wetland Quantity and Quality Using a Probabilistic Sampling Design in the Redwood River Watershed, Minnesota, U.S.A." *Wetlands* 28 (2): 324–35.

Gernes, Mark C., and Judy C. Helgen. 2002. "Indexes of Biological Integrity (IBI) for Large Depressional Wetlands in Minnesota." MPCA Final Report to US EPA # CD-995525-01. St. Paul, MN.

———. 1997. "Problem Investigation: Deformed Frogs in Granite Falls, MN, in 1993 and 1994." Minnesota Pollution Control Agency. Report to US EPA under grant no. X995749-01. St. Paul, MN.

Gilbertson, Michael, T. Kubiak, J. Ludwig, and G. Fox. 1991. "Great Lakes Embryo Mortality, Edema, and Deformities Syndrome (GLEMEDS) in Colonial Fish-eating Birds: Similarity to Chick-edema Disease." *Journal of Toxicology and Environmental Health* 33:455–520.

Gillilland, Merritt G., III, and P. M. Muzzall. 2002. "Amphibians, Trematodes, and Deformities: An Overview from Southern Michigan." *Comparative Parasitology* 69 (1): 81–85.

Gilliom, Robert J., J. E. Barbash, C. G. Crawford, P. A. Hamilton, J. D. Martin, N. Nakagaki, L. H. Nowell, J. C. Scott, P. E. Stackelberg, G. P. Thelin, and D. M. Wolock. 2007. *The Quality of Our Nation's Waters: Pesticides in the Nation's Streams and Ground Water, 1991–2001.* USGS Circular 1291. Reston, VA: USGS.

Gough, Michael. 1998. "Environmental Science and Sound Science." *Regulation* 21 (1/ Winter): 13–14. Washington, DC: Cato Institute.

Gross, Joel M., and Lynn Dodge. 2005. *Clean Water Act: Basic Practice Series.* Chicago: American Bar Association Publishing.

Grow, Doug. 2001. "State Sorely Lacking in Green Leadership." *Minneapolis Star Tribune,* Dec. 23, 02B.

———. 1998. "State's Watchdog Losing Its Bark, as Well as Its Bite: Pollution Control

Agency Has Bought into Business Mantra: 'Regulation Bad.'" *Minneapolis Star Tribune,* Dec. 14, 02.

Guillete, Louis J., and T. M. Edwards. 2005. "Is Nitrate an Ecologically Relevant Endocrine Disrupter in Vertebrates?" *Integrative and Comparative Biology* 45:19–27.

Guwande, Atul. 1999. "The Cancer-Cluster Myth." *The New Yorker,* Feb. 8:34.

Harr, Jonathan. 1995. *A Civil Action.* New York: Vintage.

Harremoës, Poul, D. Gee, M. MacGarvin, A. Stirling, J. Keys, B. Wynne, and S. Guedes Vaz. 2002. *The Precautionary Principle in the 20th Century: Late Lessons from Early Warnings.* Sterling, VA: Earthscan Publications Ltd.

Harris, Tom. 1991. *Death in the Marsh.* Washington, DC: Island Press.

Hayes, Tyrone B. 2004. "There Is No Denying This: Defusing the Confusion About Atrazine." *BioScience* 54:1138–49.

Hayes, Tyrone B., L. L. Anderson, V. R. Beasley, S. R. de Solla, T. Iguchi, H. Ingraham, P. Kestemont, J. Kniewald, Z. Kniewald, V. S. Langlois, E. H. Luque, K. A. McCoy, M. Munoz-de-Toro, T. Oka, C. A Oliveira, F. Orton, S. Ruby, M. Suzawa, L. E. Tavera-Mendoza, V. L. Trudeau et al. 2011. "Demasculinization and Feminization of Male Gonads by Atrazine: Consistent Effects across Vertebrate Classes." *Journal of Steroid Biochemistry and Molecular Biology* 107 (1–2): 64–73.

Hayes, Tyrone B., V. Khoury, A. Narayan, M. Nazir, A. Park, T. Brown, L. Adame, E. Chan, D. Buchholz, T. Stueve, and S. Gallipeau. 2010. "Atrazine Induces Complete Feminization and Chemical Castration in Male African Clawed Frogs (*Xenopus laevis*)." *Proceedings of the National Academy of Sciences* 107 (10): 4612–17.

Hayes, Tyrone B., K. Haston, M. Tsue, A. Hoans, C. Haeffele, and A. Vont. 2003. "Atrazine-induced Hermaphroditism at 0.1 ppb in American Leopard Frogs (*Rana pipiens*): Laboratory and Field Evidence." *Environmental Health Perspectives* 111:568–75.

Hayes, Tyrone B., A. Collins, M. Lee, M. Mendoza, N. Noriega, A. A. Stuart, and A. Vonk. 2002. "Hermaphroditic, Demasculinized Frogs after Exposure to the Herbicide Atrazine at Low, Ecologically Relevant Doses." *Proceedings of the National Academy of Sciences USA* 99 (8): 5476–80.

Helgen, Judy C. 2002a. *Developing an Invertebrate Index of Biological Integrity for Wetlands.* Methods for Evaluating Wetlands Condition # 9. EPA-822-R-02-019. Washington, DC: US EPA Office of Water.

———. 2002b. *The Macroinvertebrate Index of Biological Integrity (IBI). A Citizen's Guide to Biological Assessment of Wetlands. Field and Laboratory Protocols, Pictorial Keys to Wetland Invertebrates.* St. Paul: Minnesota Pollution Control Agency.

———. 1997. "The Frogs of Granite Falls: Frogs as Biological Indicators." In *Minnesota's Amphibians and Reptiles: Their Conservation and Status.* Moriarty, John J. and Delvin Jones, eds. 55–57. Lanesboro, MN.: Serpent's Tale Natural History Book Distributors.

———. 1996. Interview by host Steve Curwood on National Public Radio's weekly science program, *Living on Earth.* Oct. 11.

———. 1992. *The Biology and Chemistry of Waste Stabilization Ponds in Minnesota.* St. Paul: Minnesota Pollution Control Agency.

Helgen, Judy C., and Mark C. Gernes. 2001. "Monitoring the Condition of Wetlands:

Indexes of Biological Integrity Using Invertebrates and Vegetation." In *Bioassessment and Management of North American Freshwater Wetlands,* edited by Russell B. Rader, Darold P. Batzer, and Scott A. Wissinger, 167–85. New York: John Wiley and Sons.

Helgen, Judy C., Mark C. Gernes, Dorothy Bowers, Robert G. McKinnell, David M. Hoppe, Joel Chirhart, Jon Haferman, and Jeff Canfield. 2000. "Field Investigations of Malformed Frogs in Minnesota 1993–1997." In Kaiser et al., eds., *Investigating Amphibian Declines,* 96–112.

Helgen, Judy C., Mark C. Gernes, David Hoppe, Robert G. McKinnell, and Debra L. Carlson. 1998. "Investigation of Deformed Frogs in Minnesota 1996." A Report to the Legislative Commission on Minnesota Resources from the Minnesota Pollution Control Agency, St. Paul, MN.

Helgen, Judy C., Robert G. McKinnell, Mark C. Gernes. 1998. "Investigation of Malformed Northern Leopard Frogs in Minnesota." In *Status and Conservation of Midwestern Amphibians,* edited by Michael J. Lannoo, 288–300. Iowa City: University of Iowa Press.

Hemming, J. M., and B. Starkel. 2010. *Preliminary Assessment for Abnormal Amphibians on the National Wildlife Refuges in the Southeast Region. FY 2009.* Southeast Regional Office of US Fish and Wildlife Service. Atlanta, GA.

Honsowetz, D. C. 1996. "State Agency Wants to Hear Only the PC Explanation for Freak Frogs." *Minneapolis Star Tribune,* Nov. 12.

Hoppe, David M. 2011. Personal communication to author.

———. 2000. "History of Minnesota Frog Abnormalities: Do Recent Findings Represent a New Phenomenon?" In Kaiser et al., eds., *Investigating Amphibian Declines,* 86–89.

Hoppe, David M., and Robert G. McKinnell. 1997. "Observations on the Status of Minnesota Leopard Frog Populations." In *Minnesota's Amphibians and Reptiles, Their Conservation and Status,* edited by John J. Moriarty and Delvin Jones, 38–42. Lanesboro, MN: Serpent's Tale Natural History Book Distributors.

Hornung, Michael W., S. J. Degitz, L. M. Korte, J. M. Olson, P. A. Kosain, A. L. Linnum, and J. E. Tietge. 2010. "Inhibition of Thyroid Hormone Release from Cultured Amphibian Thyroid Glands by Methimazole, 6-Propylthiouracil, and Perchlorate." *Toxicological Sciences* 118 (1): 42–51.

Hurley, P. M, R. N. Hill, and R. J. Whiting. 1998. "Mode of Carcinogenic Action of Pesticides Inducing Thyroid Follicular Cell Tumors in Rodents." *Environmental Health Perspectives* 106 (8): 437–45.

Iyer, Poorni. 2001. "Developmental and Reproductive Toxicity of Pesticides." In vol. 1, *Principles,* 378–423. *Handbook of Pesticide Toxicology,* edited by Robert I. Krieger. 2nd ed. New York: Academic Press.

Johnson, Pieter T. J., M. K. Reeves, S. K. Krest, and A. E. Pinkney. 2010. "A Decade of Deformities: Advances in Our Understanding of Amphibian Malformations and Their Implications." In Sparling et al., eds., *Ecotoxicology of Amphibians and Reptiles,* 511–36.

Johnson, Pieter T. J., and J. Bowerman. 2010. "Do Predators Cause Frog Deformities? The Need for an Eco-epidemiological Approach." *Journal of Experimental Zoology, Part B: Molecular and Developmental Evolution* 314B (7): 515–18.

Johnson, Pieter T. J., and Richard B. Hartson. 2009. "All Hosts Are Not Equal:

Explaining Differential Patterns of Malformations in an Amphibian Community." *Journal of Animal Ecology* 78:191–201.

Johnson, Pieter T. J., J. M. Chase, K. L. Dosch, J. Gross, R. B. Hartson, D. Larson, D. R. Sutherland, and S. R. Carpenter. 2007. "Aquatic Eutrophication Promotes Pathogenic Infection in Amphibians." *Proceedings of the National Academy of Sciences* 104:15781–86.

Johnson, Pieter T. J., K. B. Lunde, D. A. Zelmer, and J. K. Werner. 2003. "Limb Deformities as an Emerging Parasitic Disease in Amphibians: Evidence from Museum Specimens and Resurvey Data." *Conservation Biology* 17 (6): 1724–37.

Johnson, Pieter T. J., K. B. Lunde, E. M. Thurman, E. G. Ritchie, S. W. Wray, D. R. Sutherland, J. M. Kapfer, T. J. Frest, J. Bowerman, and A. R. Blaustein. 2002. "Parasite (*Ribeiroia ondatrae*) Infection Linked to Amphibian Malformations in the Western United States." *Ecological Monographs* 72:151–68.

Johnson, Pieter T. J., K. B. Lunde, E. G. Ritchie, and A. E. Launer. 1999. "The Effect of Trematode Infection on Amphibian Limb Development and Survivorship." *Science* 284:802–4.

Jones, Perry M., E. M. Thurman, E. Little, S. Kersten, J. Helgen, and B. Scribner. 1999. "Pesticide and Metabolite Concentrations in Sediments, and Surface and Ground Water Found at Sites Where Frog Malformations Are Present in Minnesota." Unpublished summary report. USGS. Mounds View, MN.

Jooste, Alarik M., L. H. Du Preez, J. A. Carr, J. P. Giesy, T. S. Gross, R. J. Kendall, E. E. Smith, G. L. Van Der Kraak, and K. R. Solomon. 2005. "Gonadal Development of Larval Male *Xenopus laevis* Exposed to Atrazine in Outdoor Mesocosms." *Environmental Science and Technology* 39 (14): 5255–61.

Kadokami, Kiwao, and Masayoshi Takeishi. "Report of Deformed Frogs (Extra Forelimbs) in Japan." Undated report of 1995 findings in Yamada Park in Kitakyushu City in southern Japan. Personal communication to the author.

Kaiser, H., G. S. Casper, and N. Bernstein, eds. 2000. *Investigating Amphibian Declines: Proceedings of the 1998 Midwest Declining Amphibians Conference.* Special issue, *Journal of the Iowa Academy of Science* 107 (3, 4).

Karasov, William H., R. E. Jung, S. Vanden Langenberg, and T. Bergeson. 2005. "Field Exposure of Frog Embryos and Tadpoles along a Pollution Gradient in the Fox River and Green Bay Ecosystem in Wisconsin." *Environmental Toxicology and Chemistry* 24 (4): 942–53.

Karr, James R., and Ellen W. Chu. 1999. *Restoring Life in Running Waters: Better Biological Monitoring.* Washington, DC: Island Press.

Kashiwagi, K., N. Furuno, S. Kitamura, S. Ohta, K. Sugihara, K. Utsumi, H. Hanada, K. Taniguchi, K. Suzuki, and A. Kashiwagi. 2009. "Disruption of Thyroid Hormone Function by Environmental Pollutants." *Journal of Health Science* 55 (2): 147–60.

Kawano, S., C. Sasaki et al. 2000. *Nakaikemi, a Miraculous Lowland Marsh in Central Honshu, Japan: A Search for the Secret of Its Fascination.* The Nakaikemi Wetland Trust, Naturalist Tsuruga, Nakaikemi Women's Conservation Group, Nakaikemi Shibora Club, Biodiversity Defense Network, Japan.

Keddy, Paul A. 2010. *Wetland Ecology: Principles and Conservation.* 2nd ed. Cambridge: Cambridge University Press.

Kiesecker, Joseph M. 2002. "Synergism Between Trematode Infection and Pesticide

Exposure: A Link to Amphibian Limb Deformities in Nature?" *Proceedings of the National Academy of Sciences* 99 (15): 9900–9904.

Kloas, Werner, I. Lutz, T. Springer, H. Krueger, J. Wolf, L. Holden, and A. Hosmer. 2009. "Does Atrazine Influence Larval Development and Sexual Differentiation in *Xenopus laevis*?" *Toxicological Sciences* 107 (2): 376–84.

Kolpin, D. W., J. K. Stamer, D. A. Goolsby, and E. M. Thurman. 1998. "Herbicides in Ground Water of the Midwest: A Regional Study of Shallow Aquifers, 1991–1994." USGS Fact Sheet 076-98. Reston, VA: USGS.

Kubiak, Tim, H. J. Harris, L. M. Smith, T. R. Schwartz, D. L. Stalling, J. A. Trick, L. Sileo, D. E. Docherty, and T. C. Erdman. 1989. "Microcontaminants and Reproductive Impairment of the Forster's Tern on Green Bay, Lake Michigan—1983." *Archives of Environmental Contamination and Toxicology* 18:706–27.

Landis, Ben Young. April 18, 2011. "Endangered Frog Eggs Released to Wild Stream." USGS Amphibian Research Monitoring Initiative. News and Stories. http://armi.usgs.gov/story/story.php.

Lannoo, Michael J. In press. "A Perspective on Amphibian Conservation in the United States." *Alytes.*

———. 2011a. Personal communication to author.

———. 2011b. "No Retreat, Baby, No Surrender." *Herpetological Reviews* 42 (2): 142–45.

Lannoo, Michael J. 2008. *Malformed Frogs.* Berkeley: University of California Press.

———, ed. 2005. *Amphibian Declines: Conservation Status of United States Species.* Berkeley: University of California Press.

———, ed. 1998. *The Status and Conservation of Midwestern Amphibians.* Iowa City: University of Iowa Press.

Lannoo, Michael J., D. R. Sutherland, J. P. Rosenberry, D. Klaver, D. M. Hoppe, P. T. J. Johnson, K. B. Lunde, C. Facemire, and J. M. Kapfer. 2003. "Multiple Causes for the Malformed Frog Phenomenon." In *Multiple Stressor Effects in Relation to Declining Amphibian Populations,* edited by G. Linder, E. Little, S. Krest, and D. Sparling, 233–62. Special Publication 1443. West Conshohocken, PA: ASTM International.

Lawler, Joshua J., J. E. Aukema, J. B. Grant, B. S. Halpern, P. Kareiva, C. R. Nelson, K. Ohleth, J. D. Olden, M. A. Schlaepfer, B. R. Silliman, and P. Zaradic. 2006. "Conservation Science: A 20-year Report Card." *Frontiers in Ecology and Environment* 4 (9): 473–80.

Lee, Kathy E., V. S. Blazer, N. D. Denslow, R. M. Goldstein, and P. J. Talmage. 2000. *Use of Biological Characteristics of Common Carp (Cyprinus carpio) To Indicate Exposure to Hormonally Active Agents in Selected Minnesota Streams.* USGS Water-Resources Investigations Report 00-4202. Reston, VA: USGS.

Lehman, C. M., and B. K. Williams. 2010. "Effects of Current-Use Pesticides on Amphibians." In Sparling et al., eds., *Ecotoxicology of Amphibians and Reptiles,* 167–202.

Lehrman, Sally. 1996. "Snake Parasite May Be behind Deformed Frogs." *San Francisco Examiner,* Oct. 21, A.

Lemly, A. Dennis. 2008. "Aquatic Hazard of Selenium Pollution from Coal Mining." In *Coal Mining: Research, Technology and Safety,* edited by Gerald B. Fosdyke, chap. 6, 167–83. Nova Publishers.

———. 1999. "Contaminant Impacts on Freshwater Wetlands: Kesterson National

Wildlife Refuge, California." In Lewis et al., eds., *Ecotoxicology and Risk Assessment for Wetlands,* 191–206.

Lenkowski, J. R., J. M. Reed, L. Deininger, and K. A. McLaughlin. 2008. "Perturbation of Organogenesis by the Herbicide Atrazine in the Amphibian *Xenopus laevis.*" *Environmental Health Perspectives* 116:223–30.

Levey, Richard N., D. Shambaugh, D. Fort, and J. Andrews. 2003. *Investigations into the Causes of Amphibian Malformations in the Lake Champlain Basin of New England.* Final Report. Waterbury, VT: Vermont Department of Environmental Conservation.

Lewis, Michael A., F. L. Mayer, R. L. Powell, M. K. Nelson, S. J. Klaine, M. G. Henry, and G. W. Dickson, eds. 1999. *Ecotoxicology and Risk Assessment for Wetlands.* Pensacola, FL: SETAC Press.

Lien, Dennis. 2001. "Frog Research Loses." *St. Paul Pioneer Press,* June 5, B1.

———. 1996. "Scientists Find Deformed Frogs across the State." *St. Paul Pioneer Press,* Oct. 5, B1.

Linder, Greg, C. M. Lehman, and J. R. Bidwell. 2010. "Ecotoxicology of Amphibians and Reptiles in a Nutshell." In Sparling et al., eds., *Ecotoxicology of Amphibians and Reptiles,* 69–103.

Linzey, D. W., J. Burroughs, L. Hudson, M. Marini, J. Robertson, J. Bacon, M. Nagarkatti, and P. Nagarkatti. 2003. "Role of Environmental Pollutants on Immune Functions, Parasitic Infections, and Limb Malformations in Marine Toads and Whistling Frogs from Bermuda. *International Journal of Environmental Health Research* 13:125–48.

Little, Edward E., and Robin D. Calfee. 2010. "Solar UV Radiation and Amphibians. Factors Mitigating Injury." In Sparling et al., eds., *Ecotoxicology of Amphibians and Reptiles,* 449–73.

Little, Edward E., L. Cleveland, R. Calfee, and M. G. Barron. 2000a. "Assessment of the Photoenhanced Toxicity of Weathered Oil to the Tidewater Silverside." *Environmental Toxicology and Chemistry* 19 (4): 926–32.

Little, Edward E., R. Calfee, L. Cleveland, R. A. Skinker, A. Zaga-Parkhurst, and M. C. Barron. 2000b. "Photo-enhanced Toxicity in Amphibians: Synergistic Interactions of Solar Ultraviolet Radiation and Aquatic Contaminants." In Kaiser et al., eds., *Investigating Amphibian Declines,* 67–71.

Loeffler, I. Kati, D. L. Stocum, J. F. Fallon, and C. U. Meteyer. 2001. "Leaping Lopsided: A Review of the Current Hypotheses Regarding Etiologies of Limb Malformations in Frogs." *The Anatomical Record (New Anatomist)* 265:228–45.

Losure, Mary. 2001. "MPCA Backs Away from Frog Research." Minnesota Public Radio, June 12.

Louv, Richard. 2008. *Last Child in the Woods: Saving Our Children from Nature-Deficit Disorder.* Chapel Hill, NC: Algonquin Books .

Lucier, George. 1997. "Criticism of Frog Problem in *Post* Story Rash, Self-serving, Baseless." Commentary, *St. Paul Pioneer Press,* Nov. 13.

Malchow, Sarah. 1995. "Students' Discovery at Ney Preserve Catches Interest of MPCA." *Henderson (MN) Independent* 119 (34, Aug. 24): 1.

Mann, Reinier M., R. V. Hyne, C. B. Choung, and S. P. Wilson. 2009. "Amphibians and Agricultural Chemicals: Review of Risks in a Complex Environment." *Environmental Pollution* 157:2903–27.

McCallum, Malcolm L., and S. E. Trauth. 2003. "A Forty-three Year Museum Study of Northern Cricket Frog (*Acris crepitans*) Abnormalities in Arkansas: Upward Trends and Distributions." *Journal of Wildlife Diseases* 39 (3): 522–28.

McCormick, Tori J. 2007. "The Driftless Area. Coldwater Trout Streams Call Attention to the Ecological Health of Southeastern Minnesota." *MN Conservation Volunteer,* Mar./Apr.:8–17. St. Paul: MN Department of Natural Resources.

McCoy, Krista A., L. J. Bortnick, C. M. Campbell, H. J. Hamlin, L. J. Guillette Jr., and C. M. St. Mary. 2008. "Agriculture Alters Gonadal Form and Function in the Toad *Bufo marinus.*" *Environmental Health Perspectives* 116 (11): 1526–32.

Meersman, Tom. 2008. "Ag Giants Persuade MPCA to Alter Rule." *Minneapolis Star Tribune,* Jan 21, 1A.

———. 2007a. "Fired Scientist Says State Muzzled Him." *Minneapolis Star Tribune,* June 20, 1B.

———. 2007b. "Lovely But 'Impaired'—Never Before Has the State Declared Streams Seriously Polluted from Pesticides." *Minneapolis Star Tribune,* July 16, 1A.

———. 2007c. "Fired State Hydrologist Claims Pesticide Risk." *Minneapolis Star Tribune,* Oct. 11, 09B.

———. 2001a. "Cutback in Frog Studies Has Scientists Dismayed." *Minneapolis Star Tribune,* Jan. 19, 1B.

———. 2001b. "Ventura's Budget Has Less for Deformed-Frog Research." *Minneapolis Star Tribune,* Jan. 29, 1B.

———. 2001c. "MPCA Seen as Adrift, in Disarray." *Minneapolis Star Tribune,* Apr. 22, 1B.

———. 2001d. "State's Investigation of Deformed Frogs Is at a Standstill." *Minneapolis Star Tribune,* May 29, 1A.

———. 2001e. "MPCA, Expecting Budget Shortfall, Prepares to Retool Again." *Minneapolis Star Tribune,* June 27, 1B.

———. 1997a. "State Plans To Reduce Research on Frogs." *Minneapolis Star Tribune,* Jan. 15, 1B.

———. 1997b. "MPCA Defends Pullback on Frog Research." *Minneapolis Star Tribune,* Jan. 16, 1B.

———. 1997c. "Private Wells Linked to Frog Deformities." *Minneapolis Star Tribune,* Oct. 1, 1A.

———. 1997d. "Lab Tests on Deformed Frogs Faulty, EPA Charges." *Minneapolis Star Tribune,* Nov. 5, 1B.

———. 1997e. "Lab in Duluth Adds New Fuel to Debate on Deformed Frogs." *Minneapolis Star Tribune,* Nov. 17, 1A.

Merrell, David J. 1969. "Natural Selection in a Leopard Frog Population." *Journal of the Minnesota Academy of Science* 35 (2–3): 86–89.

Meteyer, Carol U. 2000. *Field Guide to Malformations of Frogs and Toads with Radiographic Interpretations.* Biological Science Report. USGS/BRD/BSR-2000-0005. Reston, VA: USGS.

Meteyer, Carol U., I. K. Loeffler, J. F. Fallon, K. A. Converse, E. Green, J. C. Helgen, S. Kersten, R. Levey, L. Eaton-Poole, and J. G. Burkhart. 2000. "Hind Limb Malformations in Free-living Northern Leopard Frogs (*Rana pipiens*) from Maine, Minnesota, and Vermont Suggest Multiple Etiologies." *Teratology* 62:151–71.

Meyer, M. T., and E. M. Thurman. 1996. "Herbicide Metabolites in Surface Water and Ground Water." *American Chemical Society Symposium* Series 630.

Minnesota Pollution Control Agency (MPCA). 2006. *A Comprehensive Wetland Assessment, Monitoring and Mapping Strategy for Minnesota.* St. Paul, MN.

Minnesota Administrative Rules. Chapter 4717 (Environmental Health Rules). St. Paul, MN: Office of the Revisor of Statutes. https://www.resisor.mn.gov/rules.

———. Chapter 7050 (Water Quality Standards) St. Paul, MN: Office of the Revisor of Statutes. https://www.revisor.mn.gov/rules.

———. Chapter 8420.0100 (Board of Water and Soil Resources, Wetlands Conservation Act). St. Paul, MN: Office of the Revisor of Statutes.

Mizgireuv, I. V., N. L. Flax, L. J. Borkin, and V. V. Khudoley. 1984. "Dysplastic Lesions and Abnormalities in Amphibians Associated with Environmental Conditions." *Neoplasm* 31 (2): 175–81.

Mitsch, William J., and James G. Gosselink. 2007. *Wetlands.* 4th ed. New York: John Wiley and Sons.

Mooney, Chris. 2005. *The Republican War on Science.* New York: Basic Books.

Mooney, Chris, and Sheril Kirshenbaum. 2009. *Unscientific America: How Scientific Illiteracy Threatens Our Future.* New York: Basic Books.

Munger, Mark. 2009. *Mr. Environment: The Willard Munger Story.* Duluth, MN: Cloquet River Press.

National Research Council (Committee on Mitigating Wetland Losses). 2001. *Compensating for Wetland Losses under the Clean Water Act.* Washington, DC: Water Science and Technology Board. www.nap.edu/catalog.

Ochoa-Acuna, Hugo, and Cristina Carbajo. 2009. "Risk of Limb Birth Defects and Mother's Proximity to Cornfields." *Science of the Total Environment* 407:4447–51.

Ohio EPA. 2010. *2010 Integrated Water Quality Monitoring and Assessment Report.* Final Report Mar. 8. Columbus: Ohio Environmental Protection Agency.

Oldfield, Barney, and John J. Moriarty. 1994. *Amphibians and Reptiles Native to Minnesota.* Minneapolis: University of Minnesota Press.

Olson, Erik D. 1999. "Bottled Water: Pure Drink or Pure Hype?" New York: National Resources Defense Council. April.

Ouellet, Martin, J. Bonin, J. Rodriguez, J.-L. DesGranges, and S. Lair. 1997. "Hindlimb Deformities (Ectromelia, Ectrodactyly) in Free-Living Anurans from Agricultural Habitats." *Journal of Wildlife Diseases* 33 (1): 95–104.

Paganelli, Alejandra, V. Gnazzo, H. Acosta, S. L. Lopez, and A. E. Carrasco. 2010. "Glyphosate-based Herbicides Produce Teratogenic Effects on Vertebrates by Impairing Retinoic Acid Signaling." *Chemical Research in Toxicology* 23 (10): 1586–95.

Pauli, Bruce D., J. A. Perrault, and S. L. Money. 2000. *RATL: A Database of Reptile and Amphibian Toxicology Literature.* Technical Report Series no. 357. Hull, Québec, Canada: National Wildlife Research Centre.

Pinkney, A. E. (Fred), S. A. Smith, and J. Leisenring. 2007. *Investigation of Abnormalities in Frogs on the Eastern Shore of Maryland.* Publ. No. CBFO-CO6-03. November. Washington, DC: US Fish and Wildlife Service.

Pinkney, A. E. (Fred), L. Eaton-Poole, E. M. LaFiandra, K. J. Babbitt, C. M. Bridges Britton, E. E. Little, W. L. Cranor. 2006. *Investigation of Contaminant Effects on Frog Development at Great Bay National Wildlife Refuge, Newington, New Hampshire.* Project ID 1261-5N37. Washington, DC: US Fish and Wildlife Service,.

Pittman, Craig, and Matthew Waite. 2009. *Paving Paradise: Florida's Vanishing Wetlands and the Failure of No Net Loss.* Gainesville: University Press of Florida.

Presley, Jerry J. 1996. "Deformed Frogs Caused by Fluke." Letter to the editor, *St. Louis Post-Dispatch,* Nov. 16.

Quotidien. 1996. "Des Déformations Étrange Chez des Grenouille Americaines . . . Mystère." No. 415, Oct. 4. Paris, France.

Rebuffoni, Dean. 1995a. "Deformed Frogs Prompt Investigation." *Minneapolis Star Tribune,* Sept. 1.

———. 1995b. "Mutants—or What? In the Valley of the Jolly Green Giant." *Minneapolis Star Tribune,* Nov. 25.

Reeves, Mari K., P. Jensen, C. L. Dolph, M. Holyoak, and K. A. Trust. 2010. "Multiple Stressors and the Cause of Amphibian Abnormalities." *Ecological Monographs* 80 (3): 423–40.

Reeves, Mari K., C. L. Dolph, H. Zimmer, R. S. Tjeerdema, and K. A. Trust. 2008. "Road Proximity Increases Risk of Skeletal Abnormalities in Wood Frogs from National Wildlife Refuges in Alaska." *Environmental Health Perspectives* 116 (8): 1009–14.

Reeves, Mari K., and Kimberly A. Trust. 2008. *Contaminants as Contributing Factors to Wood Frog Abnormalities on the Kenai National Wildlife Refuge, Alaska.* Final Report. Technical Paper AFWFO TR #2008-01. Washington, DC: US Fish and Wildlife Service.

Relyea, Rick A. 2005. "The Lethal Impact of Roundup® on Aquatic and Terrestrial Amphibians." *Ecological Applications* 15 (4): 1118–24.

Relyea, Rick A., and D. K. Jones. 2009. "The Toxicity of Roundup® Original Max to 13 Species of Larval Amphibians." *Environmental Toxicology and Chemistry* 28 (9): 2004–8.

Rice, Timothy M., B. J. Blackstone, W. L. Nixdorf, and D. H. Taylor. 1999. "Exposure to Lead Induces Hypoxia-like Responses in Bullfrog Larvae (*Rana catesbeiana*)." *Environmental Toxicology and Chemistry* 18 (10): 2283–88.

Rigert, Joe, and Chris Ison. 1998. "State Watchdog Lags in Policing Water Pollution." *Minneapolis Star Tribune,* Dec. 13.

Rohr, Jason R., A. M. Schotthoefer, T. R. Raffel, H. J. Carrick, N. Halstead, J. T. Hoverman, C. M. Johnson, L. B. Johnson, C. Lieske, M. D. Piwoni, P. K. Schoff, and V. R. Beasley. 2008a. "Agrochemicals Increase Trematode Infections in a Declining Amphibian Species." *Nature* 455:1235–39 and online supplements.

Rohr, Jason R., T. R. Raffel, S. K. Sessions, and P. J. Hudson. 2008b. "Understanding the Net Effects of Pesticides on Amphibian Trematode Infections." *Ecological Applications* 18 (7):1743–53.

Rollins-Smith, Louise A. 1998. "Metamorphosis and the Amphibian Immune System." *Immunological Reviews* 166:221–30.

Rollins-Smith, Louise A., and D. C. Woodhams. 2011b. "Amphibian Immunity: Staying in Tune with the Environment." In *Ecoimmunology,* edited by Gregory E. Demas and R. J. Nelson. Oxford: Oxford University Press.

Rollins-Smith, Louise A., J. P. Ramsey, J. D. Pask, L. K. Reinert, and D. C. Woodhams. 2011a. "Amphibian Immune Defenses Against Chytridiomycosis: Impacts of Changing Environments." 1–11. *Integrative and Comparative Biology.*

Rollins-Smith, Louise A., B. D. Hopkins, and L. K. Reinert. 2004. "An Amphibian Model to Test the Effects of Xenobiotic Chemicals on Development of the Hematopoietic System." *Environmental Toxicology and Chemistry* 23 (12): 2863–67.

Rorabaugh, Jason C. 2005. "*Rana pipiens* Schreber, 1782." In *Amphibian Declines: The Conservation Status of United States Species,* edited by Michael J. Lannoo, 570–77. Berkeley: University of California Press.

Sack, Steve. 1998. "News Item: Deformed Frog Cases Increasing . . ." Editorial cartoon, *Minneapolis Star Tribune,* Mar. 29.

———. 1997. "My Uncle Ole from Minnesota—He'll Be Staying with Us . . ." Editorial cartoon, *Minneapolis Star Tribune,* Oct. 6.

Sacramento Bee. 1996. "Biologist Finds Deformed Frogs in Nevada County." Oct. 6, B1.

Schardein, James L. 2000. *Chemically Induced Birth Defects.* New York: Marcel Dekker.

Schotthoefer, Anna M., M. G. Bolek, R. A. Cole, and V. R. Beasley. 2009. "Parasites of the Mink Frog (*Rana septentrionalis*) from Minnesota, U.S.A." *Comparative Parasitology* 76 (2): 240–46.

Schotthoefer, Anna M., A. V. Koehler, C. U. Meteyer, and R. A. Cole. 2003. "Influence of *Ribeiroia ondatrae* (Trematoda: Digenea) Infection on Limb Development and Survival of Northern Leopard Frogs (*Rana pipiens*): Effects of Host Stage and Parasite-Exposure level." *Canadian Journal of Zoology* 81:1144–53.

Schreinemachers, Dina M. 2003. "Birth Malformations and Other Adverse Perinatal Outcomes in Four U.S. Wheat-producing States." *Environmental Health Perspectives* 111 (9): 1259–64.

Schmid, William. 1982. "Survival of Frogs in Low Temperatures." *Science* 215:697–98.

Sessions, Stanley K. Nov., 1996. "Evidence Supports Theory That Frog Deformities Caused by Parasites." Commentary, *Outdoor News, the Sportsman's Weekly* (Plymouth, MN):1, 9.

Sessions, Stanley K., and B. Ballengée. 2010. "Explanations for Deformed Frogs: Plenty of Research Left To Do (A Response to Skelly and Benard)." *Journal of Experimental Zoology, Part B: Molecular and Developmental Evolution* 314B:341–46.

Sessions, Stanley K., and S. B. Ruth. 1990. "Explanations for Naturally Occurring Supernumerary Limbs in Amphibians." *Journal of Experimental Zoology* 254:38–47.

Shulman, Seth. 2006. *Undermining Science: Suppression and Distortion in the Bush Administration.* Berkeley: University of California Press.

Silver, Beth. 1996. "Deformed Frogs Found in Region." Associated Press. *Rapid City (SD) Journal,* Oct. 10, A6.

Skelly, David K., and M. F. Benard. 2010. "Mystery Unsolved: Missing Limbs in Deformed Amphibians." *Journal of Experimental Zoology, Part B: Molecular and Developmental Evolution* 314B:179–81.

Skelly, David K., S. R. Bolden, and K. B. Dion. 2010. "Intersex Frogs Concentrated in Suburban and Urban Landscapes." *EcoHealth* 7:374–79.

Skelly, David K., S. R. Bolden, L. K. Freidenburg, N. A. Freidenfelds, and R. Levey. 2007. "*Ribeiroia* Infection Is Not Responsible for Vermont Amphibian Deformities." *EcoHealth* 4:156–63.

Souder, William. 2000. *A Plague of Frogs: The Horrifying True Story.* New York: Hyperion.

———. 1997. "Colleagues Say Frog Deformity Researchers Leaped Too Soon." *Washington Post,* Nov. 3, 1.

———. 1996. "In Minnesota Lakes, an Alarming Mystery." *Washington Post,* Sept. 30, 1.

Sparling, Donald W. 2010a. "Ecotoxicology of Organic Contaminants to Amphibians." In Sparling et al., eds., *Ecotoxicology of Amphibians and Reptiles,* 261–88.

Sparling, Donald W., G. Linder, C. A. Bishop, and S. K. Krest, eds. 2010. *Ecotoxicology of Amphibians and Reptiles.* 2nd ed. Boca Raton, FL: SETAC CRC Press.

Steeger, Thomas, M. Frankenberry, L. Eisenhauer, and J. Tietge. 2007. *White Paper on the Potential for Atrazine to Affect Amphibian Gonadal Development.* Washington DC: EPA Office of Prevention, Pesticides, and Toxic Substances, Office of Pesticide Programs.

Stuart, Simon N., J. S. Chanson, and N. A. Cox. 2004. "Status and Trends of Amphibian Declines and Extinctions Worldwide." *Science* 306:1783–86.

Sutherland, Dan R., J. M. Kapfer, M. J. Lannoo, and M. G. Knutson. 2002. "Role of *Ribeiroia ondatrae* (Platyhelminthes: Trematoda) Metacercariae in the Development of Malformed Frogs in Minnesota and Wisconsin." In *Farm Ponds as Critical Habitats for Native Amphibians.* Final Report. Submitted to the Legislative Commission on Minnesota Resources by USGS Upper Midwest Environmental Sciences Center, La Crosse, Wisconsin.

Sweeney, P., and J. Caple. 1999. "'The Dean' an Eternal Legacy: Willard Munger 1911–1999." *St. Paul Pioneer Press,* July 12.

Takeishi, Masayoshi. 2011. Personal communication to author of 2010 findings with photographs of malformed frogs and survey data from Kitakyushu Museum of Natural History and Human History, Kitakyusu City, Japan.

———. 1996. "On the Frog, *Rana ornativentris,* with Supernumerary Limbs Found at Yamada Greenery Area in Kitakyushu City, Fukuoka Prefecture, Japan." *Bulletin of the Kitakyushu Museum of Natural History* 15 (Mar. 28):119–41.

Thurman, E. Michael. 2003. "Discovery of Chlorinated Degradates of Cyanazines in Minnesota Ground Water [abs]." *Proceedings of 2003 Spring Specialty Conference on Agricultural Hydrology and Water Quality.* Technical Publication Series no. TPS-03-1. Middleburg, VA: American Water Resources Association.

Thurman, E. Michael, D. A. Goolsby, M. T. Meyer, M. S. Mills, M. L. Pomes, and D. W. Kolpin. 1992. "A Reconnaissance Study of Herbicides and Their Metabolites in Surface Water of the Midwestern United States Using Immunoassay and Gas Chromatography/Mass Spectrometry." *Environmental Science and Technology* 26:2440–47.

Tietge, Joseph E., B. C. Butterworth, J. T. Haselman, G. W. Holcombe, M. W. Hornung, J. J. Korte, P. A. Kosian, M. Wolfe, and S. J. Degitz. 2010. "Early Temporal Effects of Three Thyroid Hormone Synthesis Inhibitors in *Xenopus laevis.*" *Aquatic Toxicology* 98 (1): 44–50.

Tietge, Joseph E, G. W. Holcombe, K. M. Flynn, P. A. Kosian, J. J. Korte, L. E. Anderson, D. C. Wolf, and S. J. Degitz. 2005. "Metamorphic Inhibition of *Xenopus laevis* by Sodium Perchlorate: Effects on Development and Thyroid Histology." *Environmental Toxicology and Chemistry* 24:926–33.

Tietge, Joseph E., S. A. Diamond, G. T. Ankley, D. L. DeFoe, B. W. Holcombe, K. M. Jensen, S. J. Degitz, G. E. Elonen, E. Hammer. 2001. "Ambient Solar UV Radiation Causes Mortality in Larvae of Three Species of *Rana* under Controlled Exposure Conditions." *Photochemistry and Photobiology* 74 (2): 261–68.

Turpen, J. B. 1998. "Induction and Early Development of the Hematopoietic and Immune Systems in *Xenopus.*" *Developmental and Comparative Immunology* 22 (3): 265–78.

Union of Concerned Scientists. 2008. *Interference at the EPA. Science and Politics at the U.S. Environmental Protection Agency.* Cambridge, MA: Union of Concerned Scientists.

Uhlenbrock, T. 1996. "5-Legged Frog Crops up in Missouri Pond." *St. Louis Post-Dispatch,* Oct. 20.

US Army Corps of Engineers 1987. *Wetlands Delineation Manual.* Wetlands Research Program Technical Report Y-87-1 (online edition). Environment Laboratory, Waterways Experiment Station. Vicksburg, MS. www.wetlands.com/regs/tlpge02e.htm.

US EPA. 2009a. *Chemicals Evaluated for Carcinogenic Potential.* Washington, DC: EPA Office of Pesticide Programs.

———. 2009b. *Six-Year Review Health Effects Assessment: Summary Report.* EPA 822-R-09-006. Washington, DC: EPA Office of Science and Technology, Office of Water (4304T).

———. 2000. *Cyanazine; Cancellation Order.* Federal Register. 65 (4): 771–73. EPA. OPP-300000/46C; FRL-6486-7. January 6. Washington, DC. [See also Exotoxnet.orst.edu for US EPA actions on the herbicide cyanazine.]

———. 1972. Clean Water Act (CWA), 101a. Also Sect. 404 of the Act. PL 92-500.

———. *Wetland Bioassessment Publications: Wetlands Modules.* http://water.epa.gov/scitech/swguidance/standards/criteria/nutrients/wetlands/index.cfm#modules.

US Fish and Wildlife Service. 2011. "Assessment of Amphibian Abnormalities on U.S. National Wildlife Refuges: Ten Year Summary Report (2000–2009)." Arlington, VA: US Fish and Wildlife Service.

US Geological Survey. 2000. Mercury in the Environment. USGS Fact Sheet 146-00. October.

———. 1998. *Herbicides in Ground Water of the Midwest: A Regional Study of Shallow Aquifers, 1991–94.* USGS Fact Sheet 076-98. July. USGS Kansas Water Center. Reston, VA: USGS.

US Government Accountability Office. 2008. *Chemical Assessments. Low Productivity and New Interagency Review Process Limit the Usefulness and Credibility of EPA's Integrated Risk Assessment Information System.* GAO-08-440. Washington, DC: US Government Accountability Office.

US House of Representatives. 2003. *Politics and Science in the Bush Administration.* Committee on Government Reform—Minority Staff. Prepared for Rep. Henry A. Waxman. August 2003. Updated Nov. 2003. www.reform.house.gov/min.

Vanden Langenberg, Sue M., J. T. Canfield, and J. A. Magner. 2003. "A Regional Survey of Malformed Frogs in Minnesota (USA)." *Environmental Monitoring and Assessment* 82:45–61.

Vileisis, Ann. 1997. *Discovering the Unknown Landscape: A History of America's Wetlands.* Washington, DC: Island Press.

Walker, Matt. 2009. "Legless Frogs Mystery Solved." BBC News. *Earth News.* June 25.

Ward, Mary H., B. A. Kilfoy, P. J. Weyer, K. E. Anderson, A. R. Folsom, and J. R. Cerhan. 2010. "Nitrate Intake and the Risk of Thyroid Cancer and Thyroid Disease." *Epidemiology* 21 (3): 389–95.

Weinhold, Bob. 2009. "Environmental Factors in Birth Defects." *Environmental Health Perspectives* 117 (10): A441–47.

Weir, Linda, I. J. Fiske, and J. A. Royle. 2009. "Trends in Anuran Occupancy from Northeastern States of the North American Amphibian Monitoring Program (NAAMP)." *Herpetological Conservation and Biology* 4 (3): 389–402.

Weiss, Rick. 2004. "'Data Quality' Law is Nemesis of Regulation." *Washington Post,* Aug. 16, A01.

Wellock, Thomas R. 2007. *Preserving the Nation: The Conservation and Environmental Movements. 1870–2000.* Wheeling, IL: Harlan Davidson.

Werner, Earl E., K. L. Yurewicz, D. K. Skelly, and R. A. Relyea. 2007. "Turnover in an Amphibian Community: The Role of Local and Regional Factors." *Oikos* 116:1713–25.

Weselak, M., T. E. Arbuckle, and W. Foster. 2007. "Pesticide Exposure and Developmental Outcomes: the Epidemiological Evidence." *Journal of Toxicology and Environmental Health, Part B* 10:41–80.

Wilson, Greg A., T. L. Fulton, K. Kendell, G. Scrimgeour, C. Paszkowski, and D. W. Coltman. 2008. "Genetic Diversity and Structure in Canadian Northern Leopard Frog (*Rana pipiens*) Populations: Implications for Reintroduction Programs." *Canadian Journal of Zoology* 86 (8): 863–74.

Winchester, Paul D., J. Huskins, and J. Ying. 2009. "Agrichemicals in Surface Water and Birth Defects in the United States." *Acta Paediatrica* 98:664–69.

Zaga, Angela, E. E. Little, C. F. Rabeni, and M. R. Ellersieck. 1998. "Photoenhanced Toxicity of a Carbamate Insecticide to Early Life Stage Anuran Amphibians." *Environmental Toxicology and Chemistry* 17:2543–53.

Zhang, F., S. J. Degitz, G. W. Holcombe, P. A. Kosian, J. E. Tietge, N. Veldhoen, and C. C. Helbing. 2006. "Evaluation of Gene Expression Endpoints in the Context of a *Xenopus laevis* Metamorphosis-Based Bioassay to Detect Thyroid Hormone Disruptors." *Aquatic Toxicology* 76:24–36.

Ziegler, D. 1996. "Deformed Frogs Popping up All over Minnesota: A Ribbit-ing Discovery." *Boca Raton (FL) National Examiner,* Dec. 17.

INDEX

American toad, 76, 205, 214
amphibian decline, 2, 148, 158, 201, 212, 218, 222
aquatic invertebrates, 9, 56, 68, 72, 167, 179. See also *Daphnia;* dragonflies
atrazine: and frogs, 160, 212, 215; regulation of, 26, 55, 216; in water, 126, 167, 187, 215. *See also* pesticides

biological monitoring: of aquatic invertebrates, 69, 121, 179, 167; and biological integrity, 3, 9, 29; Index of Biological Integrity (IBI), 121, 191; testing polluted waters, 3, 9, 191; of wetlands, 4, 9, 71–72, 220
bird deformities, 135, 181
birth defects, 23, 53, 89, 98, 211; and agriculture, 24, 62–63, 112, 211–12. *See also* retinoic acid
bone defects, 101, 111–14, 136, 208, 217. *See also* deformity types; lathyrogenic chemicals

cancer, 24–26, 43, 112, 159–60
Carson, Rachel, 2–4, 29
chytrid fungus, 156–58, 201, 218
Clean Water Act, 3–4, 29, 127, 189–94; impaired waters listings, 191, 220; water quality standards, 30, 120, 127, 210–11, 220; and wetlands, 189–94, 218–20

Daphnia, 36, 52, 68–70, 167
Data Quality Act, 107, 192
deformed frogs, causes of, 31, 135–36, 180, 186, 207, 210; chemicals, 13–14, 31, 38, 50, 53, 56, 88–89, 110–12, 186–87, 213–14; parasites, 53, 88–90, 101, 114–15, 204–8; predation, 208–9; ultraviolet light, 30–31, 89, 143–48, 213–14; in water, 161–63, 170, 178, 186, 213
deformed frogs, global, 87, 155–56
deformed frogs, Minnesota: citizens' reports, 82–85, 90–91, 172, 202; in Granite Falls, 7–8; in Henderson, 7, 10, 17; in Litchfield, 20–26, 60; at Ney Pond, 15, 17, 24, 44, 48–49, 56, 94, 112, 187, 207
deformed frogs, surveys of: background rates, 84–85, 202–3; increasing trends, 106, 115, 202–3; national reporting, 172, 203–4; species of frogs, 87–88; USFWS surveys, 203–6
deformity types: bent limbs, bony triangles, 83, 88, 196, 203, 205–6, 213; eye missing or misplaced, 17, 82, 87–88, 114, 161, 203; internal, 114, 208; jaw, head malformed, 88, 114, 161, 203; limb misplaced, 5, 85, 149, 206, 208; limb missing or partial, 24, 53, 82–85, 104, 202, 205–6, 208; multiple, branched limb or foot, 7, 11, 15–18, 53,

deformity types (*continued*)
82, 85, 88–90, 102–3, 203–6, 214; percentage of types, 202–3
dragonflies: as clean water indicators, 121, 209; life cycle, 69–71; as predators, 208–9

endocrine-disrupting chemicals, 51, 69–71, 79, 88, 159–60, 211, 217
EPA. *See* US EPA

Freedom of Information Act (FOIA), 118–20
frog calls, 74–76, 223
frog embryo assay, 52–53, 162, 164, 168–69
frog evolution, 76, 218, 221–22
frog life cycles: development, 88–89, 114, 161–64, 213–15; eggs, egg masses, 85, 137, 141–48, 185, 198; migrations and habitats, 21, 44–45, 201; overwintering, 44–48

Gernes, Mark, 11–12, 36

Hoppe, David, 21, 84, 87, 149
hormones and frogs, 185–86, 216–17. *See also* endocrine-disrupting chemicals; thyroid hormone

immune systems, frog and human, 186, 205, 213, 215, 218
impaired waters listings, EPA, 191, 220
Index of Biological Integrity (IBI), 121, 191. *See also* biological monitoring
isolated wetlands. *See under* wetlands

Kesterson disaster, 13–14, 44. *See also* bird deformities; selenium

lathyrogenic chemicals, 111–15, 136
leopard frog. See *Rana pipiens*
Lithobates pipiens. See *Rana pipiens*
Little, Edward (USGS), 143–44, 148, 222

malformed frogs. *See* deformed frogs
McKinnell, Robert, 21, 23, 38, 40, 48–49
media coverage of frogs, 17, 20, 66, 95–103, 118, 163–65
metals, 50, 103, 178, 212–13
mink frog (*Rana septentrionalis*), 87, 149–51, 207
Minnesota frog investigation: field study design, 83, 91, 141; state budget issues, 38–39, 60–63, 88, 120–23, 129–33, 177, 188–90, 195. *See also* deformed frogs, surveys of
Minnesota frogs. *See* deformed frogs, Minnesota
Minnesota New Country School, 11, 30, 59, 65–66
Minnesota Pollution Control Agency (MPCA), 39, 54, 124–28, 133–34, 188–92, 195, 210, 220. *See also* Minnesota frog investigation
Munger, Willard, 28–29, 34, 40–41, 59–62, 123, 131, 177–78, 194

National Institute of Environmental Health Sciences (NIEHS), 109–10, 113, 139, 152, 161, 173, 180, 186, 213, 217
National Wildlife Health Center Lab (USGS), 92, 114, 150, 180, 206, 222
Ney Pond. *See under* deformed frogs, Minnesota
NIEHS. *See* National Institute of Environmental Health Sciences
nitrates, 52, 217. *See also* thyroid hormone
northern leopard frog. See *Rana pipiens*

parasites, 16, 53, 88, 90, 101, 114–15, 161, 164, 204–7, 213–15
pesticides: and amphibians, 54, 88, 110, 151, 159, 213–16; breakdown products, 52, 56–58, 112; cyanazine, 112, 187, 212; glyphosate, 157, 214–15; and humans, 26, 62–63, 211–12; "inert" ingredients, 56–57; malathion, 26, 214; methoprene,

50, 89; in Minnesota waters, 56, 89, 111, 126, 187; in Ney Pond, 187; in rainfall, 55, 89, 187; regulation, 26, 29, 54, 56, 210–11; in rural wells, 55, 167, 210; usage not public, 57. *See also* atrazine
polluted waters listings. *See* impaired waters listings
predation and deformities. *See* deformed frogs, causes

Rana pipiens (leopard frog), 21, 203–4; declining in U.S., 40, 185, 201, 215; prevalence in surveys, 149, 203. *See also* frog life cycles
Rana septentrionalis. See mink frog
Reinitz, Cindy, 7, 11, 45, 64, 66
retinoic acid, 53, 88–89, 110, 214
Ribeiroia, 205–7, 214. *See also* parasites

science and government, 45, 118–20, 125–28, 132, 175, 179–80, 192, 195, 216. *See also* Data Quality Act; Freedom of Information Act
selenium, 13–14, 38, 213. *See also* bird deformities; Kesterson disaster
Straub, Art, 46–47, 222–23
student testimonies, 30, 46, 60–61, 64–67

teachers: Reinitz, Cindy, 7, 11, 45, 64, 66; Straub, Art, 46–47, 222–23; Thovson, Gail, 20–21, 60–61
teratogens. *See* bird deformities; birth defects; lathyrogenic chemicals; retinoic acid
Thousand Friends of Frogs, 59, 63, 100
Thovson, Gail, 20–21, 60–61
thyroid hormone, 43, 51–52, 160, 173, 186, 216–17

ultraviolet light, 89, 158; and frog eggs, 30–31, 137, 145, 213; and genetic damage, 144; penetrance in ponds, 143–48; photochemical conversions, 89, 144, 214; and photolyase repair, 144, 148
US EPA, 3–4, 68–69; deformed frog investigations, 8, 137, 142, 151; funding wetlands research, 8–9, 12, 36–37, 39, 179, 189, 193; regulating water pollution, 26, 29–30, 52, 55–56, 71–72, 112, 127, 187, 210, 215, 217; regulating wetlands, 189–90. *See also* Clean Water Act
US Fish and Wildlife Service (USFWS), 32; frog surveys in refuges, 203–4
US Geological Survey (USGS): deformed frog investigations, 163, 180, 187, 213; function and history, 143–44; and pesticides, 55–56, 188–89, 212, 215. *See also* Little, Edward; National Wildlife Health Center Lab

volunteer monitoring: deformed frog surveys, 172; frog call surveys, 72–75, 223; wetland monitoring, 72–74, 79–80, 179, 199, 223

water quality standards, 3, 30, 54, 127, 165, 170, 210, 220. *See also* Clean Water Act
well water tests, 162–63, 167. *See also* pesticides: in rural wells
Wetland Health Evaluation Project (WHEP). *See* volunteer monitoring
wetlands: as amphibian habitat, 44–45, 86; definition, 32; isolated wetlands, 5, 45, 218–21; losses, 32, 86, 219; pollution, 10, 33, 70, 72, 81, 187, 207; regulation, 189–90, 218–20. *See also* biological monitoring; Clean Water Act
World Congress of Herpetology (Prague), 154–60

Xenopus. See frog embryo assay

zooplankton. See *Daphnia*

JUDY HELGEN grew up in Massachusetts, graduated from Mount Holyoke College with a major in zoology, and earned a master's degree from Columbia University. After marrying and raising two sons, she reentered graduate school, earning a PhD in zoology at the University of Minnesota. By then her interests had shifted from molecular biology to aquatic ecology and understanding how pollution affects aquatic organisms. As a research scientist at the Minnesota Pollution Control Agency, she worked to promote biological monitoring of wetlands and became the agency's lead investigator into the widespread occurrences of deformed frogs. She has taught at Metropolitan State University and St. Olaf College, is a member of the Society of Environmental Journalists, and now lives in Roseville, Minnesota.